Acclaim for **Kary Mullis's**

DANCING NAKED IN THE MIND FIELD

"This bona-fide wild card of the scientific community writes with eccentric gusto. . . . Mullis has created a free-wheeling adventure yarn that just happens to be the story of his life." —*Elle*

"Entertaining . . . [Mullis is] usefully cranky and combative, raising provocative questions about received truths from the scientific establishment."
—*The New Yorker*

"A very good book by a fascinating man. . . . [Mullis] enjoys an almost frighteningly brilliant mind and yet somehow manages to keep his feet firmly on the ground. . . . This guy cuts through the nonsense to the quick, tells it like it is, and manages to do so with insouciance [and] occasional puckishness. . . . But lighter moments aside, what he has to say is important." —F. Lee Bailey

Kary Mullis

DANCING NAKED IN
THE MIND FIELD

Kary Mullis was born in North Carolina in
1944. He grew up in Columbia, South Carolina, and attended the Georgia Institute of
Technology. He received his doctorate in biochemistry in 1973 from the University of California at Berkeley. In 1993 Dr. Mullis was
awarded the Nobel Prize and the Japan Prize
for his invention of the polymerase chain reaction (PCR). He lives in Newport Beach,
California, with his wife, Nancy.

DANCING NAKED IN
THE MIND FIELD

DANCING NAKED IN
THE MIND FIELD

Kary Mullis

Vintage Books

A Division of Random House, Inc.

New York

FIRST VINTAGE BOOKS EDITION, JANUARY 2000

The Library of Congress has cataloged
the Pantheon edition as follows:
Mullis, Kary B.
Dancing naked in the mind field / Kary Mullis.
p.   cm.
Includes bibliographical references and index.
ISBN 0-679-44255-3
1. Science—Miscellanea—Popular works.   II. Title.
Q173.M96   1998
081 – dc21   97-48879
CIP

Vintage ISBN: 978-0-679-77400-6

www.vintagebooks.com

Printed in the United States of America

I dedicate this book to Nancy Lier Cosgrove Mullis.

Jean-Paul Sartre somewhere observed that we each of us make our own hell out of the people around us. Had Jean-Paul known Nancy, he may have noted that at least one man, someday, might get very lucky, and make his own heaven out of one of the people around him. She will be his morning and his evening star, shining with the brightest and the softest light in his heaven. She will be the end of his wanderings, and their love will arouse the daffodils in the spring to follow the crocuses and precede the irises. Their faith in one another will be deeper than time and their eternal spirit will be seamless once again.

Or maybe he would have just said, "If I'd had a woman like that, my books would not have been about despair."

This book is not about despair. It is about a little bit of a lot of things, and, if not a single one of them is wet with sadness, it is not due to my lack of depth; it is due to a year of Nancy, and the prospect of never again being without her.

CONTENTS

DANCING NAKED IN
THE MIND FIELD

1

THE INVENTION OF PCR

THE SUN HAD been hot that day in Mendocino County. It was May. A dry wind had come out of the east, and nobody knew how hot it had been until, around sunset, the wind stopped. I drove up from Berkeley through Cloverdale headed to Anderson Valley. The California buckeyes poked heavy blossoms out into Highway 128. The pink and white stalks hanging down into my headlights looked cold, but they were loaded with warmed oils that dominated the dimension of smell. It seemed to be the night of the buckeyes, but something else was stirring.

My little silver Honda's front tires pulled us through the mountains. My hands felt the road and the turns. My mind drifted back into the lab. DNA chains coiled and floated. Lurid blue and pink images of electric molecules injected themselves somewhere between the mountain road and my eyes.

I see the lights on the trees, but most of me is watching something else unfolding. I'm engaging in my favorite pastime.

Tonight, I am cooking. The enzymes and the chemicals I have at Cetus are my ingredients. I am a big kid with a new car and a full tank of gas. I have shoes that fit. I have a woman sleeping next to me and an exciting problem, a big one that is in the wind.

"What subtle cleverness can I devise tonight to read the sequence of the King of molecules?"

DNA. The big one.

There are pressing reasons to want to read this molecule. Children are born with genetic defects, sometimes with tragic consequences like muscles that wither and die. Such things could be predicted and averted if we could read the DNA blueprints.

Then there are the looming but not so pressing reasons for knowing DNA—the ones that extend out to horizons mankind has not yet reached. Understanding the intricate mechanisms of our own genes will have more than just medical impact. It will be one of the long and tangled strands of our future as a growing civilization of the planet Earth. Having a detailed understanding of why children resemble parents will lead to genetic manipulating by those who may prefer alterations over strict duplication. Genetic engineering will not be a new endeavor. Evolution is and always has been a genetic engineer. It's just that people with eyes and minds and imagination see things in the distance and they get impatient. They will want control and they will want it soon and they will have it. The DNA molecules in our cells are our history, and they are the stuff of which our future will be crafted. All of the organs of all of the plants and animals of Earth and organs that have never been in the light of the moon or the sun, will be ours to explore—to use and adapt to our needs. Our will be done on Earth as we sail off to the stars in heaven.

Yes, DNA is the big one. Tonight I am playing with a fire that will burn as brightly as Antares, descended behind these fragrant mountains several hours ago.

The key to the problem lies in the oligonucleotides that my laboratory at Cetus now easily makes. Like a "FIND" sequence in a computer search, a short string of nucleotides in a synthetic molecule might be able to define a position along a very much longer natural DNA molecule. Finding a place to start is of utmost importance. Natural DNA is a tractless coil, like an unwound and tangled audio tape on the floor of the car in the dark.

What kind of chemical program would be required to "FIND" a specific sequence on DNA with 3 billion nucleotides and then display that sequence to a human who was trillions of times larger than the DNA? Instead of a list of statements in BASIC or FORTRAN run on a computer and displayed on a screen, I had to arrange a series of chemical reactions, the result of which would represent and display the sequence of a stretch of DNA. The odds were long. Like reading a particular license plate out on Interstate 5 at night from the moon.

I knew computer programming, and from that I understood the power of a reiterative mathematical procedure. That's where you apply some process to a starting number to obtain a new number, and then you apply the same process to the new number, and so on. If the process is multiplication by two, then the result of many cycles is an exponential increase in the value of the original number: 2 becomes 4 becomes 8 becomes 16 becomes 32 and so on.

If I could arrange for a short synthetic piece of DNA to find a particular sequence and then start a process whereby that sequence would reproduce itself over and over, then I would be close to solving my problem.

The concept was not out of the question because in fact one of the natural functions of DNA molecules is to reproduce themselves. They do it every time a cell divides into two daughter cells. A short piece of synthetic DNA could be treated in such a way that it would stick to a longer strand of DNA in a specific way if the sequences matched up somewhere on the long piece. The matching process would not be perfect. I might locate a thousand different places that were similar to the one I was searching for in addition to the correct one. A thousand out of the 3 billion in the human genome would be no trivial feat, but it wouldn't be enough. I needed to find just one place.

Suddenly, I knew how to do it. If I could locate a thousand sequences out of billions with one short piece of DNA, I could use another short piece to narrow the search. This one would be designed to bind to a sequence just down the chain from the first sequence I had found. It would scan over the thousand possibilities out of the first search to find just the one I wanted. And using the natural properties of DNA to replicate itself under certain conditions that I could provide, I could make that sequence of DNA between the sites where the two short search strings landed reproduce the hell out of itself. In one replicative cycle I could have two copies, and in two cycles I could have four, and in ten cycles. . . . I thought I remembered that two to the tenth was around a thousand.

"Holy shit!" I hissed and let off the accelerator. The car coasted into a downhill turn. I pulled off. A giant buckeye stuck out from the hill. It rubbed against the window where

Jennifer, my girlfriend and co-worker, was asleep, and she stirred. I found an envelope and a pencil in the glove compartment. Jennifer wanted to get moving. I told her something incredible had just occurred to me. She yawned and leaned against the window to go back to sleep.

We were at mile marker 46.58 on Highway 128, and we were at the very edge of the dawn of the age of PCR. I could feel it. I wrote hastily and broke the lead. Then I found a pen.

I confirmed that 2^{10} was 1,024. I must have smiled. If I repeated this new reaction ten times, I'd get a thousand copies of some piece of DNA, any piece of DNA, the molecule that knew everything about everything. Twenty cycles would give me a million, thirty would give me a billion. I could still smell the buckeyes, but they were drifting a long way off. I pulled back onto the highway, and Jennifer made a sound of approval that we were under way again. She had no idea where we were headed.

About a mile down the canyon, I pulled off again. The thing had just exploded again. A new and wonderful possibility. Not only could I make a zillion copies, but they would always be the same size. That was important. That was the almighty, the halleluja! clincher. The hell with Jennifer. I had just solved the two major problems in DNA chemistry. Abundance and distinction. And I had done it in one stroke. I stopped the car at a nice comfortable turnout and took my time working my way through the consequences. This simple technique would make as many copies as I wanted of any DNA sequence I chose, and everybody on Earth who cared about DNA would want to use it. It would spread into every biology lab in the world.

I would be famous. I would get the Nobel Prize.

"Ten years from now they will know me in Zambia," I boldly conjectured, "and Alice Springs."

"Ten years from now I'll walk into biochemical laboratories at the University of East Jesus, and they'll ask me to say something wise to the graduate students."

Somehow, I thought, it had to be an illusion. It was too easy. Someone else would have done it and I would surely have heard of it. We would be doing it all the time. What was I failing to see? "Jennifer, wake up."

She wouldn't wake up. I had thought of incredible things before that somehow lost some of their sheen in the light of day. This one could wait until morning. But I didn't sleep that night. We got to my cabin and I started drawing little diagrams on every horizontal surface that would take pen, pencil, or crayon, until dawn when, with the aid of a bottle of good Anderson Valley Cabernet, I settled into a perplexed semi-consciousness.

Afternoon came—including new bottles of celebratory red fluids from Jack's Valley Store—but I was still puzzled, alternating between being absolutely pleased with my good luck and clever brain and being mildly annoyed at myself and Jennifer for not seeing the flaw that must have been there. I had no phone at the cabin and there were no other biochemists besides Jennifer and me in Anderson Valley. The conundrum, which lingered throughout the weekend and created an unprecedented desire in me to return to work early, was compelling. If the cyclic reactions that by now were symbolized in various ways all over the cabin really worked, why had I never heard of them being used? If they had been used, I surely would have heard about it and so would everybody else, including Jennifer, who was presently sunning herself beside the pond, taking no interest in the explosions that were rocking my brain.

Why wouldn't these reactions work?

Monday morning I was in the library. The moment of truth. By afternoon it was clear. For whatever reasons, there was nothing in the abstracted literature about succeeding or failing to amplify DNA by the repeated reciprocal extension of two primers hybridized to the separate strands of a particular DNA sequence. By the end of the week, I had talked to enough molecular biologists to know that I wasn't missing anything really obvious. No one could recall such a process ever having been tried.

However, shocking to me, not one of my friends or colleagues would get excited over the potential for such a process. True, I was always having wild ideas, and this one maybe looked no different from last week's. But it *was* different. There was not a single unknown in the scheme. Every step involved had been done already. Everyone agreed that you could extend a primer on a single-stranded DNA template. Everyone knew that it would make a double-stranded DNA molecule that you could heat up and turn into two more single-stranded DNA templates. Everyone agreed that what you could do once, you could do again. Most people didn't like to do things over and over, me in particular. If I had to do a calculation twice, I preferred to write a program instead. But no one thought it was impossible. It could be done, and there was always automation. The result on paper was so obviously fantastic that even I had little irrational lapses of faith that it would really work in a tube. Most everyone who could take a moment to talk about it with me felt compelled to come up with some reason why it wouldn't work. It was not easy in that postcloning, pre-PCR year to accept the fact that you could have all the DNA you wanted. And that it would be easy.

I had a directory full of untested ideas in the computer. I opened a new file and named this one "polymerase chain reaction." I didn't immediately try an experiment, but all summer I kept talking to people in and out of the company. I described the concept around August at an in-house seminar. Every Cetus scientist had to give a talk twice a year. But no one had to listen. Most of the talks were dry descriptions of labor performed, and most of the scientists left early without comment.

One or two technicians were interested, and on the days when she still loved me, Jennifer thought it might work. On the increasingly numerous days when she hated me, my ideas and I together suffered her scorn.

I continued to talk about it and by late summer had a plan to amplify a 400-nucleotide fragment from Human Nerve Growth Factor, which Genentech had cloned and just published an article on in *Nature*. It would be dramatic. What had taken Genentech months to obtain, I would reproduce in hours.

My friend Ron Cook, who had founded Biosearch and produced the first successful commercial DNA synthesis machine, was the only person I remember during the summer who shared my enthusiasm for the reaction. He knew it would be good for the oligonucleotide business. Maybe that's why he believed in it. Or maybe he's a rational chemist with an intact brain. He's one of my best friends, so I have to disqualify myself from claiming any really objective judgment regarding him. Perhaps I should have followed his advice, but then things would have worked out differently and I probably wouldn't be here on the beach in La Jolla writing this, which I enjoy. Maybe I would be rich in Tahiti. He suggested one night at his house that since no one at Cetus had taken it seriously, I should resign my job, wait a little while, make it work, write a patent, and get rich. By rich he wasn't imagining the $300 million that Cetus finally got from Hoffmann–La Roche for PCR. Maybe one or two. The famous chemist Albert Hofmann was at Ron's that night. He had invented LSD in 1943. At the time he didn't realize what he had done. It only slowly dawned on him. And then things worked their way out over the years as no one would have ever predicted, or could have controlled by forethought and reason. Kind of like PCR.

I responded weakly to Ron's suggestion. I had already described the idea at Cetus, and if it turned out to be commercially successful, they would have lawyers after me forever. Ron was not sure that Cetus even had rights to my ideas unless they were directly related to my duties. I wasn't sure about the law, but I was pretty happy working at Cetus and assumed, innocently, that if the reaction worked big time, I would be amply rewarded by my employer. I was plenty wrong on that one.

The subject of PCR was not yet party conversation, even among biochemists, and it was quickly dropped. Albert's being there was much more interesting—even to me.

My problems with Jennifer were not getting any better. That night was no exception to the trend. I drove home alone feeling sad and unsettled—not in the mood for leaving my job or any other big change in what was left of stability in my life. PCR seemed distant and small compared to our very empty house.

In September I did my first experiment. I like to try the easiest possibilities first. So one night I put human DNA and the Human Nerve Growth Factor primers in a little screw-cap tube with an O-ring and a purple top. I boiled for a few minutes, cooled, added DNA polymerase, closed the tube and left it at 37 degrees. It was exactly midnight on the ninth of September. I poured a cold Becks into a 400-milliliter beaker and contemplated my notebook for a few minutes before leaving the lab.

Driving home, I figured that the reaction would proceed by itself, and I didn't really care how long it took as long as nobody had to do anything. For a reaction with the potential of this one—especially in the light of the absence of anything else that could do the same thing—time was a very secondary consideration. Would it work at all? The next most important thing was, Would it be easy to do? Then came, How long would it take?

At noon the next day I went to the lab to take a twelve-hour sample. There was no sign by ethidium bromide of any 400-nucleotide fragment. I could have waited another hundred years, as I had no idea what the absolute rate might be, but I succumbed slowly to the notion that I couldn't escape much longer the unpleasant prospect of cycling the reaction by hand. This meant adding the thermally unstable polymerase after every cycle and a hell of a lot more work for me.

For three months I did sporadic experiments while my life with Jennifer, at home and in the lab, was crumbling. Progress in the lab was slow. Finally I retreated from the idea of starting with human DNA. I settled on something simpler, called a plasmid. The first successful experiment happened on

December 16, 1983. It was dark outside when I took the auto-radiogram out of the freezer and developed it. There, just where it should have been, was a little black band. A tiny little black band. It meant that I was going to be famous. I remember the date. It was the birthday of Cynthia, my former wife from Kansas City, who had encouraged me to write fiction and bore us two fine sons. I had strayed from Cynthia eventually to spend two tumultuous years with Jennifer. When I was sad for any other reason, I would also grieve for Cynthia. There is a general place in your brain, I think, reserved for "melancholy of relationships past." It grows and prospers as life progresses, forcing you finally, against your better judgment, to listen to country music.

And now as December threatened Christmas, Jennifer, that crazy, wonderful woman chemist, had finally and dramatically left our house and the lab and headed for New York and her mother, for reasons that seemed to have everything to do with me but that I couldn't fathom. I was beginning to learn trag-edy. It differs a great deal from pathos, which you can learn from books. Tragedy is personal. It would add strength to my character and depth, someday, to my writing. Just right then, I would have preferred a warm friend to cook with. Hold the tragedy lessons. December is a rotten month to be study-ing your love life from a distance.

I celebrated my victory with Fred Faloona, a young mathe-matician and a wizard of many talents whom I had hired as a technician. Fred had helped me that afternoon set up this first successful PCR reaction, and I stopped by his house on the way home. As he had learned all the biochemistry he knew directly from me, he wasn't certain whether to believe me when I informed him that we had just changed the rules in molecular biology.

"Okay, Doc, if you say so." He knew I was more concerned with my life than with those cute little purple-topped tubes.

In Berkeley it drizzles in the winter. Avocados ripen at odd times, and the tree in Fred's front yard was wet and sagging from a load of fruit. I was sagging as I walked out to my little Honda Civic, which never failed to start. Neither Fred, empty Becks bottles, or the sweet smell of the dawn of the age of PCR could replace Jenny. I was lonesome.

2

THE BIG PRIZES

IN DECEMBER OF 1992 I had agreed to head a project
based on a concept of mine that, if it worked, would change
the world of medical diagnostics. Two German pharmaceutical companies had agreed to provide our company, Atomic
Tags, with $6 million to start working on it in California. They
would expect me to continue directing it. I figured it would
take ten years, with a staff of a hundred chemists, physicists,
and physicians. The research would cost maybe $30 million a
year once we got up and running, but if it worked, it would be
well worth it. Unfortunately, I would be responsible for it. You
foolish person, I thought as I flew back to California, you've let
these two businessmen talk you into making them a bunch of
money. But I had committed myself.

Waiting for me when I returned home was a letter from
the Japanese Ministry of Technology informing me that I
had been awarded the Japan Prize. It was a lot of yen. It was
1:30 A.M. and I had no idea what the exchange rate was
for the yen. I spent the night wondering whether I was rich.
In the morning I found out that I could probably live for a
while on it.

The Germans dropped a bomb just two weeks later on my

ten-year research project. The Bundestadt had passed a new law that diminished the profits of the German pharmaceutical companies by about 30 percent. That cut out any new foreign research. Atomic Tags turned belly up and I was free.

With a light heart, I was off to Japan to meet the emperor and the empress. I might be the only person ever to address the empress of Japan as "sweetie." She was kind enough not to chastise me for it. I had fun talking to her. I asked her how many other empresses she knew. She said there were only three in the world. I asked her who the other two were, and we agreed that they were not potential girl friends. "So you don't have any girl friends?"

She said without hesitation, "No girl friends."

And then she told me about her life. At first, just after the war, when her father-in-law Hirohito had been defeated and her husband had been reified into a figurehead, she had been new to the ways of the Imperial Court. She had been a commoner. She spoke out right away about things, and it caused problems. She was warned not to be herself. She learned to control her behavior. She controlled it so well that she began to act like a real empress. She became a prisoner of her position. But she did it. She learned it so well that the Japanese press, years later, was now criticizing her for being too imperial, for acting too much like an empress. She was a lovely, elegant lady.

"What does an empress do for fun?" I asked. "Can you go shopping or to a movie?"

Her life was completely organized and planned, she said. She had very little decision-making power. When I suggested several books I thought she might enjoy, she explained that someone would have to read them first to determine whether they were appropriate.

I couldn't believe that. "I'll send them directly to you."

"I don't know whether you can send me mail," she replied. "Everything must come through channels."

The empress and I really enjoyed the evening. At one point she asked me about my children, and I pointed across the room to my son, Christopher, and said, "He speaks Japanese. He was pretty shocked to learn that the American ambassador to Japan doesn't speak it at all." She was intrigued with the fact that my son spoke Japanese and suggested I bring him to a cocktail party following the reception. I couldn't reach him during the banquet. Our table was surrounded by security people assigned to keep us separate. I told her she could get anything she wanted because she was the empress. "Go ahead," I said. "Just crook your finger and bring that guy right over here." She giggled. But she did it and the security agent came right over to her. She thought that was a great discovery.

CHRISTOPHER WAS SETTLING down to some Japanese television when the knock on the door came. It was the imperial security forces and they wanted him downstairs. He dressed and came down to the cocktail party with gray-suited men on either side of him. I spied him in the doorway looking interested but also like a high school student who had been dragged away from the television. He was promptly sent through the receiving line, and the emperor's face lit up when Chris introduced himself in Japanese. It was a memorable night.

I was confident I was going to receive the Nobel Prize in 1992. The host of a German TV show had called and explained that each year he did a documentary about the winner of the Nobel Prize in chemistry, and he was preparing the 1992 show. In the past, he had successfully picked every winner of the prize for chemistry. He claimed he was a very good guesser, but I figured this bastard must get inside information, he must be getting the word from somebody on the committee. That means I'm going to win it this year. His TV crew spent a week filming me in La Jolla and Mendocino. I was very excited. And I was actively humble.

As it turned out, I had good reason to be humble. I didn't win. I stopped speculating about when I might get it and I tried not to pay attention. About six months before the 1993 awards were to be announced, my mentor from Berkeley, Joe Neilands, from whom I had learned a little bit about chemistry and a whole lot about life, told me, "I wouldn't be surprised if you got the Nobel Prize this year. But you'd make it easier for the committee to give it to you if you didn't talk to the press so much. They don't have to give it to you till you're dying."

Neilands said that it was probably okay that I admitted loving surfing and women, but he thought the committee might frown on the fact that I admitted using LSD. Surfing, women, *and* LSD might be too much, he told me. They might decide to wait until I settled down in twenty or thirty years. Joe had spent a sabbatical or two at the Karolinska in Sweden and he knew the scene. We both knew I wouldn't shut up.

After being disappointed in 1992, I stopped thinking about the Nobel Prize. The German guy never called back. I wasn't even sure when the awards were to be announced. My phone rang at 6:15 A.M. on the morning of October 13, 1993. I thought I knew who it was. On both the eleventh and twelfth someone from Japan had sent me a fax at exactly that time. He thought it was my afternoon. So when the phone rang in my bedroom I stayed in bed, knowing the fax machine would eventually pick it up. Then I heard someone leaving a message on my answering machine. I heard the words "Nobel Foundation."

I leaped out of bed. I picked up the phone just as the speaker hung up. Great, I thought, I've missed the Nobel Prize call. Will they call back? Almost instantly the phone rang again. He had heard me just as he'd hung up. "Congratulations, Dr. Mullis. I am pleased to be able to announce to you that you have been awarded the Nobel Prize."

"I'll take it!" I said. I knew that they couldn't make you take it and I didn't want there to be any doubts. We talked for a minute, and I was warned to be prepared for an assault by the media, but since this was the first time I'd ever won a Nobel Prize, there was no way I could have anticipated the response. I figured maybe I'd get ten calls or something. I didn't realize how big the known world is. As soon as I hung up, I tried to call my mother in South Carolina. Coincidentally, this was her birthday and I reckoned this was a fine birthday present. But when I picked up the phone, a reporter from the AP was on the line. The phone hadn't even rung. I spoke to him for a second, then hung up, and tried again. I picked up the phone, and someone from UPI was on the line. Then somebody from a local station called. They wanted to bring a camera crew over. Then Steve Judd showed up as he usually did around seven, and I told him that I had just won the Nobel Prize. He said, "I know. I heard it on the radio. Let's go for a surf."

The local station that wanted to bring over a camera crew was still on the line. I told them that I would be available in an hour. I needed to wake up and I would be out surfing. Of course, they asked me where we were going. I looked up at Steve and we nodded agreement. I said we would be up at Thirteenth Street in Del Mar. We headed in the other direction to Tourmaline. I needed time.

Several friends joined us. When we came out of the water, a camera crew from another station was waiting. They had gone directly to my apartment and found out from a neighbor where I usually surfed. They didn't know me, and they were asking everyone who came out of the water if he was Kary Mullis. Andy Dizon admitted to being me. They asked him how it felt to win the Nobel Prize. He proclaimed that it was like a dream come true. They asked him what he would be doing the rest of the day, and he turned to me and said, "Wow! I just remembered, *this* is Kary Mullis." They didn't show that on the nightly news.

By the time I got back home, my house was completely surrounded by print and broadcast reporters and camera crews. As it turned out, none of the other Nobel laureates that year were serious about surfing, and "Surfer Wins Nobel Prize" made headlines.

Friends began arriving with Champagne, and the party began. That afternoon I finally reached my mother. I wanted to tell her to stop sending me articles about DNA, since I had now won the Nobel Prize for my expertise on that subject. My mother often mailed me articles from *Reader's Digest* about advances in DNA chemistry. No matter how I tried to explain it to her, she never grasped the concept that I could have been writing those articles, that something I had invented made most of those DNA discoveries possible. She probably hoped that winning the Nobel Prize might enable me to be published someday in *Reader's Digest.*

The party continued for two days. Eventually it moved north to my place in Mendocino. Roederer Vineyards was just down the road, and no one failed to notice. I woke up late one afternoon from a dream that I was dead in a coffin. Winning the Nobel Prize can be hazardous to your health.

I invited my mother, my two sons, and a nice woman named Einhoff, whom I'd been dating for only a few weeks, to accompany me to Stockholm for the ceremony. I also took Cynthia, the mother of my two boys.

That year two Nobel Prizes in chemistry were awarded. Michael Smith, a Canadian who had demonstrated that you could change the sequence of a gene using oligonucleotides, was also honored with a Nobel Prize. He too invited his former wife, their children, and his girl friend to the ceremony. This kind of coincidence cannot be assigned a statistical probability because it happens only once.

I was informed that the proper dress for the awards ceremony was white tie. I went to an Italian tailor in La Jolla and he made me a beautiful set of white tails. About a week before I was to leave for Sweden, I saw some photographs that had been taken at the 1992 ceremony. The laureates all were in black. White tie in spring or summer means the suit is all white; white tie in the winter means a white tie with a black suit. I was out of season.

The tailor made me the proper suit and shipped it to Sweden. I had a suspicion that the white tails would not go to waste. When I hung them in my closet in a mothproof bag, I thought, "Someday I will get married in this outfit." I wore them four years later when I married Nancy Cosgrove.

The American laureates were honored at the White House on our way to Sweden. I was looking forward to meeting President Clinton and Hillary Clinton. I had a plan. I thought that if I had the opportunity to speak privately to the president, what I wanted to know was, "Did they pass that joint back to you after you didn't inhale? And didn't anybody tell you, 'Hey, Bill, that stuff's four hundred dollars an ounce'?" If he was by himself, I figured, he would have to smile. But the president simply rushed through the room. We shook hands, cameras focused; he congratulated each of us and was gone.

I did have the opportunity to speak with Hillary. At that time she was in charge of American health care. I wondered whether she really knew what she was doing. For example, did she know how the health care system worked in Australia? I had the feeling that if I asked her about it she would tell me that someone on her staff was an expert on the subject. She told me exactly how the health care system works in Australia. "Okay," I said. "How about Ireland? How does it work in Ireland?" She told me exactly how the health care system works in Ireland.

I came away thinking she was a smart woman. He's got a lot of charm and is taller than I pictured him. It's easy to understand how he got elected, but Hillary's the smart one.

December was a miserable time of year to be in Sweden. It was cold and dark all the time. I had already come down with the flu. But it was fun. The Swedish people take the awards very seriously, and I think they enjoyed me because I did and I didn't. Rather than being somber and stodgy, I believed it was a time to celebrate. I had been awarded the most extraordinary prize bestowed on scientists, and I was going to have a good time picking it up.

Every morning I'd get up and go to lunch with the faculty of some university. Then I'd give a lecture and rush back to the hotel to dress for some formal event. Most of the time I behaved myself very well. There was only one time when I almost got arrested, and that was not entirely my fault. As a present, R. B. Haynes had given me a little hand-held laser. It projects a red dot on the first solid object in its path, no matter how far away that object is. It's like pointing a very long finger.

In the long dark winter mornings of Stockholm I couldn't stop playing with it. I would sit in my window in the Grand Hotel and play with the Swedes. I'd shine the beam on the newspapers they were reading or on the sidewalk in front of them as they walked along. One morning a cab driver was smoking a cigarette, and I shined the beam directly in front of him. When he noticed it, he got up and returned to his cab, so I aimed it through the windshield onto his dashboard. I thought it was a funny thing to do until the police arrived.

I didn't know that a laser was often mounted on rifles and used to aim. I also did not know that just about a year earlier someone had been walking down a street in Stockholm, a red dot suddenly appeared on his chest, and he'd been shot by a sniper. The cab driver had seen me pointing the laser out of the third-floor window. When he told the police, they were a little dubious, explaining, "That's a Nobel laureate's suite." The three officers at the door asked courteously, "Dr. Mullis, have you been shining a red light out the window?" When I told them I had, they asked to see it. They wanted to be sure it was not attached to a rifle. I didn't blame them at all. I asked whether there was a law in Sweden against shining a red light out a window. There was no such law, they explained, but after the murder it did tend to make people nervous. I didn't use the laser again in Sweden.

My first official duty was to give the Nobel lecture. Normally, each laureate explains what he did to win the award and why he did it. It's often a complicated speech that nobody understands but everybody applauds. I decided I wanted to give a human account of what was taking place in my life when I invented PCR, rather than a technical explanation. "I'm going to try to explain how it was that I invented the polymerase chain reaction," I said. "There's a bit of it that will not easily translate into normal language. If that part wasn't of interest to more than a handful of people here, I would leave it out. What I will do instead is let you know when we get to that and also when we are done with it. Don't trouble yourself over it. It's esoteric and it's not crucial. I think you can understand what it felt like to invent PCR without following the details."

I proceeded to explain that I'd spent much of my life believing that science was fun and that my invention was little more than an extension of the things I had started doing as a child in Columbia, South Carolina. I mentioned that it had not been my intention to revolutionize the world of biochemistry when I invented PCR; PCR was a tool I created because I needed it to do an experiment. In truth, I was terribly naïve, I said, and if I had had more knowledge about what I was doing, PCR would never have been invented.

Following the official presentation of the medal, the king and queen hosted a banquet for about thirteen hundred people. We were served by waiters dressed in medieval costume.

And at that banquet each new laureate has an audience with the king and queen. Mostly, the royal couple and the new laureates engage in several moments of small talk. I did not think the king would be interested in oligonucleotides.

I took the opportunity to discuss a matter of importance. I knew that the king and queen were very popular with the Swedish people but that there were some problems with their daughter, the sixteen-year-old princess. Apparently the tabloids had written some negative things about her. "I wouldn't worry about it," I said. "She's a sixteen-year-old princess. If she's tolerable at all, she's fine. I'm sure she'll grow out of it." "In fact," I continued, "I'm so confident about that, that I'm willing to offer my son in marriage. He's just about the right age for her. And I would be happy to have him marry your daughter in exchange for a third of your kingdom."

My mother was thrilled to be in Sweden. I believe she had always expected that at least one of her sons would win a Nobel Prize, but she was most impressed with the fact that I was able to introduce her to CNN correspondent Lou Dobbs. She had had a crush on him for years, and for my mother, the most exciting part of the trip was getting to sit next to Lou Dobbs.

I gave my final lecture in the city of Malmö and boarded a hovercraft that was going to skim across the water to Copenhagen. By that time my picture had been in every paper every day for a week. As I sat down, a man wearing a big hat with a feather in it came over to me. "Dr. Mullis," he said, "the people of Sweden love you." And then in a great gesture, he took off his hat and bowed to me. The people on the boat applauded. It was a perfect ending.

3

A LAB IS JUST ANOTHER PLACE TO PLAY

EVERY NOVEMBER WHEN I was young, my mother would give my brothers and me a pile of catalogues and let us pick what we wanted for Christmas. It was in one of those catalogues that I found a Gilbert Chemistry Set. Something about tubes filled with things with exotic names intrigued me. My objective with that set was to figure out what things I might put together to cause an explosion. I discovered that whatever chemicals might be missing from the set could be bought at the local drugstore. In the 1950s in Columbia, South Carolina, it was considered okay for kids to play with weird things. We could go down to the hardware store and buy 100 feet of dynamite fuse, and the clerk would just smile and say, "What are you kids going to do? Blow up the bank?"

The first thing of any consequence that I made with my chemistry set was a substance similar to thermite. I mixed powdered aluminum, ammonium nitrate, a dash of something else, and heated it over an alcohol burner. When I pulled it away from the flames, the reaction kept going. The mixture got red hot, broke the test tube, and suddenly went *Fffffsshhoooo.* Now that, I thought—being only seven years old—was cool. I didn't know what had happened then, but I decided that science was going to be fun.

We were fortunate to have the Russians as our childhood enemies. We practiced hiding under our desks in case they had the temerity to drop a nuclear weapon on Columbia, South Carolina, during school hours. In 1957 the Russians launched the space race by putting Sputnik I into orbit around Earth. It was only twenty-three inches in diameter, but it revolutionized the American educational system. The government poured millions of dollars into science education. It was a fortuitous time to be young and in love with science.

Two years later, my friends and I were launching our own rockets in my backyard. Our goal was to see how high into the sky we could send a frog and bring it back alive, but we also wanted to create a huge blastoff and a long, colorful trail. To fuel our rocket, we began experimenting with different concentrations of potassium nitrate and sugar. I'd mix it in a tennis ball can and heat it on the charcoal grill. Every so often my mother would lean out the window and warn, "Now Kary B, don't you blow your eyes out!"

I'd respond cheerfully, "Okay, Mom, I won't."

But kids being kids, and explosives being explosives, occasionally the fuel would explode. One time it set a big tree on fire. This taught me an important lesson: Never mix explosive chemicals under a big tree.

The first chemistry lab in which I spent some time was Dreher High. Our teacher would leave the lab open in the afternoons so Al Montgomery and I could play in there. She was pleased we wanted to be there. Most of her students hated chemistry. Everything we played with would today be considered too dangerous for adults to use without federally approved supervision. But in 1960 chemicals were just bottles of stuff that no one took very seriously. It was perfectly acceptable to turn sixteen-year-old boys loose in a chemistry lab.

When I became president of the Junior Engineering Technical Society (JETS), Al and I thought it would be fun to put on a science show for the elementary schools in Columbia. The stated objective of the show was to demonstrate the basic principles of science as they had been explained by Isaac Newton in the seventeenth century.

The show consisted of a series of demonstrations. We rolled metal balls down inclined wooden troughs to illustrate how mass is accelerated in a gravitational field. We taught that a hypothesis is a guess that can be proven into a theory by doing experiments.

We had a dramatic opening. A concoction with iodine and potassium perchlorate. It began with burning some alcohol, quietly heating and concentrating the other ingredients in a porcelain dish behind a crack in the curtains. When the eerie blue alcohol flame died out, the residue would blaze up into serious pyrotechnic action for a second and then there would be sparks till it was over. It always worked before.

But when we did the show at A. C. Moore Elementary, my neighborhood grammar school, it didn't. The blue flame wavered for a second, and then the whole thing exploded. Shards of the crucible blew all over the place. Everyone sat in complete awe of chemistry.

I held my breath waiting for someone from the blood-stained front row to be carried backstage in the arms of my first-grade teacher. No one appeared. I walked tentatively onto the stage and started talking about Isaac Newton, checking out the front row for blood. I didn't see any.

When we were through, a boy with blond hair and a little blood on him came backstage with a small piece of glass that had hit him in the forehead. He wasn't bothered by it. It was as if he'd caught a baseball after a home run. I took the piece of glass from him and asked him not to tell. He was a kid from my neighborhood. I knew he'd keep quiet.

I first worked in a professional lab the summer after my high school graduation. My dad helped me get a job at Columbia Organic, a supplier of research chemicals. There was no excuse for such a company in a little town like Columbia, except for the person of Max Gergel. Max is an amazing entrepreneur, a courtly gentleman, and a wonderful storyteller. He wrote some of the stories down in *Excuse Me Sir, But Could I Interest You in a Kilo of Isopropyl Bromide?*

Max owned and operated Columbia Organic. He made about a thousand research chemicals, but he resold a lot more. My job was to go through his orders in the morning and find the cheapest supplier of the chemicals we were ordering. The same chemicals can have several different names, depending on who is using them and for what. I would translate them into all their chemical names in different languages.

One day I discovered a bizarre oversight that had gone unnoticed for years. We were buying a chemical from Fluka, a Swiss company, for one of our customers in Illinois. We were paying Fluka $100 a gram. No one in the company had noticed that we had a kilogram of the very same chemical in stock but under another name. When we would place an order with Fluka, they would turn around and order the required amount from us—at $24 a gram—and send it on to our customer. Fluka had to know what was going on but saw no reason to enlighten us at Columbia Organic.

I told Max about it. He thought it was hilarious. He called the guy from Fluka at home. They had a history together and this was just another funny story in a long line of funny stories. Max took me to lunch for the first time. I was no longer just his high school employee—I was his friend.

Sometimes we got orders for discontinued chemicals. Columbia Organic didn't make them, and other suppliers no longer kept them in stock. The summer after my freshman year at Georgia Tech, Al and I decided to try to supply Max with the chemicals that no one else had. We set up a lab in Al's garage. I made a deal with Max: we would use anything in his stockroom to make the chemicals and we would sell them to him for 60 percent of the standard price.

The thing we had failed to consider about these chemicals was that if they had been convenient to make, somebody else would have been doing it. Max probably knew that, but he wanted those compounds, and he saw in us something of himself many years earlier and he liked it. He wanted to see our business unfold.

We worked at night in Al's garage. Our first synthesis was nitrosobenzene. Max had given us a little sample left in an empty bottle from an old shipment. The solid was brown and oily. Nitrosobenzene should be white and crystalline. We decided ours would set a new standard for the industry. At the library of the University of South Carolina we looked up the procedure in *Organic Synthesis*. It sounded easy. We bought a five-gallon crock and a thick glass rod at a hardware store.

Into the crock we measured nitrobenzene, water, and ice.

We stirred and slowly added zinc, which reduced the nitro-benzene to phenylhydroxylamine. We filtered it away from the solid zinc oxide that had also formed in the reaction and added more ice. Now we poured in a measured amount of chromic acid with vigorous stirring and animated conversation. As described in *Organic Synthesis*, the phenylhydroxyl-amine turned into nitrosobenzene, which floated to the top. We filtered it out. It was brown and oily, but we had about 100 grams. At $4 a gram, we were rich.

We could have dried it and taken it to Max, and he would have paid us and sold it for $6 a gram, but we were not going to make inferior chemicals just to get by. We spent the remainder of the night purifying our product. When we were done, our crystals were white, with a very slight greenish cast in the early morning sunlight. It was the prettiest nitrosobenzene the chemical business would ever see. We had lost about 20 percent, but what we had was pure. We had made our first real chemical.

The next day we took the nitrosobenzene to Max, and had he not done conscientious things like that as a kid in the business, he would have been shocked. He was pleased to the point of adopting us both as his children forever. Chemists get emotional about other chemists because of the language they have in common and the burns on their hands.

Al and I went out to the stockroom to find the ingredients we would need for our next synthesis. There is an immense thrill when you gather together the ingredients to make something new. We put the bottles in the blue '55 Chevy. Gergel watched us leave, wishing he could cancel his day, and twenty years, and follow us home.

We were going to make some phenacyl bromide. No one

else in the world was willing to go to that trouble and deal with the hazards for the small market that existed.

Bromine is a dense red liquid. It fumes when you weigh it out, and it pours like mercury. If you get it on you, it leaves a deep scar. We poured it into the dropping funnel and very slowly let it drip into the stirred ether solution. It gets a little warm—if you drop it in too fast, the ether boils. Ice and patience help. We had plenty of ice.

The reaction gives off hydrogen bromide, a corrosive gas with a pungent odor. We were venting the gas through a wall fan into the humid southern night. Around one o'clock we took a beer break and discovered that the entire neighborhood was enveloped in a choking white cloud of hydrogen bromide. We figured the cloud would be dispersed by morning when the neighborhood awoke. Unfortunately, we had killed a large camellia bush that was growing under the fan.

We kept working. The phenacyl bromide crystallized and we started filtering it out of the ether. By that point it had begun to smell extremely pungent. I asked, "Al, what were you saying about the properties of this stuff?"

"*Org Syn* said it was a lachrymator."

"What's that?"

"It's like a tear gas."

That didn't make sense to me. "This is a solid, not a gas."

"I know, that's why I wasn't worried."

"My eyes are stinging, Al."

"Mine too. It's a fucking tear gas, even if it is a solid."

We went outside and washed our faces with the hose, but it was not all that effective. Phenacyl bromide was scarcely soluble in water. Like oil, it would sort of spread around in water, but it wouldn't wash away.

The night air was quiet, liquid and thick, a typical summer night in Columbia. Our faces were burning. The floral smells of the night—azaleas, camellias, rhododendrons, night-blooming jasmine—had been replaced by something wicked.

We ducked back into the garage and took the several hundred grams of phenacyl bromide off the filter and put it on trays that we set on the washer and dryer. The crystals drying in the trays looked perfect. We had made another organic chemical. We closed the door and left the exhaust fan on high. It was 3:00 A.M.

Exhausted, but still remembering a modicum of conscience, we taped a DO NOT ENTER sign to the door. Backing out of the driveway, I could see blackened leaves gently blowing off the camellia. Oh well, Al's mother had more camellias than she needed.

The next day, while Al and I were at our day job, his grandmother went into the garage to do the laundry. The moment she opened the door, the drying phenacyl bromide fumes hit her in the face like mace. When I arrived that evening, Al's mother looked at us as if we were grandmother abusers. Al's grandmother was not speaking to him. The good thing about gassing Al's grandmother was that if anyone had complained that we had killed a big bush, we could respond, "Well, maybe, but at least Ganny's still alive."

We moved the lab. Al's brother-in-law, Frank, had some land about twenty miles out into the country, where you could do just about whatever you wanted to do. We built a lab out of an old chicken coop and worked there for the next two summers.

At Georgia Tech I worked in a lab run by E. C. Ashby. He was interested in reductions with light-metal hydrides, which meant that he liked to work with solutions in ether that would explode on exposure to moisture. Moisture was plentiful in Atlanta in the summer. When we were done with some experimental solution of lithium aluminum hydride, we would have a problem as to where to put it. Ashby's solution was to take the well-sealed flasks home and host a party for the lab on the Fourth of July. The flasks would be floated out on his pond, and graduate students would shoot at them with a .22.

I was just an undergraduate and unaware of Ashby's Fourth of July event. I had been working with lithium aluminum hydride, and I thought it was my responsibility to get rid of it. To do that in the lab required a long tedious procedure, but outside the chemistry building there was a drain in the alley. I combined all the solutions in a single two-liter beaker, covered it with aluminum foil, and carried the beaker outside to the drain. I dumped it quickly, stepped back, and waited for the flames. Nothing happened.

I stood nearby. Two minutes passed. Maybe the ether had evaporated and the hydride had not yet combined with water to produce the hydrogen. Maybe it was dry down there. I didn't want to walk away, in case something horrible happened.

A campus cop walked up the alley. Please God, don't let him light a cigarette. He didn't. He was about twenty feet away from the drain. He looked at me suspiciously.

I noticed an outside water spigot on the building. As the cop watched, I filled the empty beaker with water and poured it down the drain. A deep red flame—the color of lithium atoms and the combustion of hydrogen—erupted from the drain. The cop looked from the drain to the spigot, back to the drain and back to the spigot. I knew he was thinking that water spigots aren't supposed to dispense liquids that blow up. I walked quickly into the building, wondering whether he would report the drain, the spigot, or the student. The chemistry building was old and labyrinthine, and once I passed under the furnace ducts and slipped into a cluttered hallway, I was safe.

THE LAB IN WHICH I learned the most about life was presided over by Joe Neilands. I think he taught me specifically three things:

1. Use a funnel when you are pouring acetone from a large bottle into a smaller one.
2. Drink tea instead of coffee.
3. Be responsible; you are a scientist.

Joe Neilands made me aware of the present-day planet. I already knew about the universe but had spent little time thinking about today and the people around me. Joe's lab studied micro-organisms that, when grown in the near absence of iron, could make things that would bind tightly to iron when it was provided. We were interested in the fact that iron is everywhere in biology and that it isn't soluble in water without some kind of carrier.

Joe Nielands was a remarkable man. His lab was a place to play with scientific tools. In most university labs, the function of graduate students is to conduct experiments and write papers that will advance the career of the professor running the lab. Joe was pleased when we did some work on iron transport, but it wasn't a requirement. He encouraged us to follow our interests.

As long as I wrote a thesis and got a degree, Joe didn't much care what else I did. He figured I would not be very successful in science because I was too interested in everything else, including women. He introduced me to visitors to the lab as his "wide-angled genius." Joe hoped I would get a good education. If he said anything at all to me on the grand scale, it was that my education was being paid for by the people and I owed them my conscientious best. From my home base in Joe's lab, I followed my own curiosity. It took me into anthropology, sociology, physics and math, and even music courses, where I could meet women. He warned me that my style of hanging out as long as I could as a graduate student would eventually result in some tightening of the departmental rules, and it did. But the university was a big place and there were so many courses to take. I audited as many classes as I could.

During my first year at Berkeley, the biochemistry department had purchased a brand-new Varian A60 nuclear magnetic resonance spectrometer. It was absolutely the coolest tool I had ever seen. It was a powerful analytic instrument that allowed you to identify all the different types of hydrogen atoms in a particular molecule, and from that it was possible to determine the molecular structure. It enabled a chemist to understand how the carbons and hydrogens and nitrogens and oxygens were all stuck together—literally to draw a picture of a molecule. They had one at Georgia Tech, but they never let undergrads even look at it. At Berkeley I helped take it out of the box.

It was a huge machine. To operate it, you sat in front of a

console surrounded by dials and switches. It was like sitting in the cockpit of a 747. By the time I'd read the manual and played with it enough to feel confident, the spring quarter had ended. I was taking the summer off, but I couldn't wait to get back to the machine in the fall.

When I returned to Berkeley in September, I raced up to the third floor to find the NMR. When I walked into the room, I saw that a gray plywood box had been padlocked over the top of the console. The machine looked as if it had been put in prison. It was difficult to believe that anyone would cover such a lovely instrument with an ugly box. Someone had staked a claim to my machine.

I discovered that the machine had been imprisoned by a postdoctoral chemist working in the lab of Dr. Dan Koshland. This chemist had used an A60 in Berkeley's chemistry department, but he was ignorant of a crucial fact. The A60 was a very sensitive instrument, and the longer it stood idle, the more it went out of whack. I'd found that if it hadn't been used on weekends, it would take more than an hour on Monday morning to properly adjust the amplitude of its three perpendicular magnetic fields. In fact, the manual clearly warned that this problem probably would be the worst on Monday mornings.

They didn't have that problem in the chemistry department, which was much larger than the biochemistry department. Consequently, their A60 was in use seven days a week, twenty-four hours a day. It never had down time to lose tune, so the chemists didn't have to learn how to align it. When the postdoc found our machine out of tune, he just assumed that the biochemists were screwing it up. This was typical of a chemist; chemists always believe they're smarter than biochemists. Of course, physicists think they're smarter than chemists, mathematicians think they're smarter than physicists, and, for a while, philosophers thought they were smarter than mathematicians, until they found out in this century that they really didn't have anything much to talk about.

In order to prevent biochemists from ruining his alignment, the postdoc had the carpentry shop construct the box to cover the dials. No one could use the machine without his permission. This did not sit well with me. I was not going to beg for permission to use it. I bought my own lock and slipped it through the hasp. There were now two locks on the machine.

I explained the situation to Neilands. He thought my response appropriate and pretty damn funny. It wasn't long before Dan Koshland came in. "Kary, could I see you for a moment?"

"Come on in, Dan," I said. "Have some tea."

Reluctantly, he sat for tea. "Did you put that lock on the NMR machine, Kary?"

"I did," I admitted. "Did you put the other one on?"

"My postdoc did," he confessed. "He thought people were messing it up."

"Well, if I had thought that, I would've called a meeting of all the people using the machine and we could have figured out what to do, instead of putting a crude-looking box over it. But I'll take mine off if you'll take yours off."

So we took both locks off the machine, and I had the janitors take the box away. As a compromise, an electronic key was installed. Everybody who needed a key got one. I hung a key on a nail behind the machine. No one failed to notice it.

I STARTED WORKING at Cetus in 1979. It was founded when Ron Cape, Pete Farley, Don Glaser, and Carl Djerrassi decided that a whole lot of money might be made in biotechnology. They didn't know about recombinant DNA yet, but they sensed something coming down the pipeline. Cetus hired me in 1979 to make oligonucleotides. It was a wonderful place to be doing biochemistry. I probably worked harder my first few years there than at any other time in my life because it was fun and because Ron Cape and Pete Farley made it even more fun. I was learning how to synthesize DNA and it fascinated me. There were all kinds of wild ideas floating around the halls, and there was absolutely no restraint in terms of imagination.

When I first started, Cetus was still a small company. For the first time in my life I had the personal privilege as a scientist of buying pretty much whatever I needed when I needed it. Without having to get permission, I could call a company and order anything costing less than five thousand dollars. If something cost more than that, it usually took me one phone call to Farley to get approval. Cetus provided scientists with every service imaginable. The concept was to eliminate as much of the routine work as possible, to allow scientists the time to do science.

As the company grew, forms were eventually distributed. Rules were instituted. Good people became involved in bad office politics. Cetus became just another business. I didn't blame Pete or Ron; they were just as helpless as the rest of us when the gray-suited middle managers came vulturing in after the public stock offering. The worst thing I remember about those days of swelling middle management was when the guy who used to empty the isotope disposal buckets became the "safety officer" and suddenly got a staff, an office, and power.

Safety officers have a vested interest in interpreting everything in terms of various degrees of danger. In order to live for another day and to develop respect for the safety officer, signs were posted everywhere reminding us that everything we did was dangerous. Enclosed with every chemical is a Materials Handling Data Safety Sheet that explains its potential hazards—it's the law. The person who wrote the instructions for sodium chloride must have thought it was a mixture of sodium metal and chlorine gas rather than a completely innocuous compound that people sprinkle on foods to enhance the taste, usually called salt. Sodium and chlorine are pretty serious separately but not when combined into sodium chloride. The safety sheet described in detail the method that should be used to clean up sodium chloride: "Wear rubber boots," it advised. "Wear a respirator. A small spill can be flooded with water. Cleaning up a larger spill may require more than one person." Since I did not want anyone in my lab to suffer serious salt injury, I posted this warning on the wall. This is what happens when government agencies, who have to answer to nobody in particular, run rampant. If you want to have sodium chloride in your lab, you must have safety equipment that would be appropriate for sodium metal and chlorine gas. If you want to have it in a restaurant, you just have to have a salt shaker.

The safety officer at Cetus and I had a real serious battle. I never called him the safety officer; I called him the danger officer because all he ever did was put up DANGER signs. A danger officer wants to find dangerous things because it gives him more power. Just like a toxicologist would like to find as many toxins as possible. If you are paid to be a safety officer in a lab, you will find danger whether there is any or not.

At Cetus the most dangerous thing was the blue punch that David Gelfand made for the Blue Death Party at the scientific retreats, which he concocted from alcohol and something blue. Many people, myself included, drank it until we did things that were so out of line with normal social discourse that we either fell over in some inconvenient place or, as in the case of what happened to me one night, got into fights with fellow researchers. The safety officer should have put some signs around Gelfand's blender. The things the safety officer did made it more difficult to work and could have resulted in more accidents. His work always reminded me of the military officer in Vietnam, who explained: "In order to save that village, it became necessary to destroy it."

I had several work cabinets with sliding hoods made of thick safety glass. They enabled people to work safely with dangerous materials. I came in one day to find that he had plastered stickers on everything. Hazardous materials. Noxious materials. Radioactive materials. Corrosive materials. Signs posted everywhere warned that safety glasses had to be worn at all times. One hood was almost half-covered with his stickers. Overnight my lab had become a very dangerous place. I went storming into his office. "What do you think those hoods are made out of glass for?" I screamed at him. "With all your stickers covering the glass, you can't see a damn thing!" When I calmed down, I explained, "My people don't need to be constantly reminded they're in danger. They're not in that much danger or we wouldn't be doing it. With all your little signs everywhere, nobody can tell what's dangerous and what's not, because according to you everything is dangerous."

With a razor blade and solvents, I scraped off all his danger stickers. Eventually we compromised. I allowed him to post those signs required by law but not on any transparent surfaces. And we agreed that every time he put up a new sign, he had to take down an old one.

The biggest battle I fought with the danger officer was over the fact that I insisted on keeping my lunch and a case of Beck's beer in the same fridge in which I kept my radioactive isotopes. I kept the beer in bottles on the bottom shelf and the radioactive isotopes in a sealed lead-lined container on the top shelf. I pointed out to him that there was no way known to science that anything, even radiation, could escape from a closed lead-lined container into a sealed bottle. "I'm planning on drinking most of that beer myself," I said. "If it wasn't safe, I wouldn't put it in there."

Fortunately, Pete Farley, the president of Cetus, liked my beer. He liked coming into my lab in the afternoon and taking a bottle from my refrigerator. This presented the danger officer with a dangerous situation, so he took the safest way out: He stopped searching my refrigerator.

Once he hired a beautiful young woman as his assistant. He sent her into my lab to conduct a safety inspection. I eventually called her *Nostradama Salutatis* or "Our Lady of Safety." The danger officer thought she might be able to handle me. Instead, I invited her home for dinner, and several months later she moved in with me.

As it turned out, the most dangerous thing in my lab was associated with Our Lady of Safety. One afternoon a man who thought he was her boyfriend kicked open my door and started threatening me. That was the only time in my life that I worried about my safety in a laboratory.

4

FEAR AND LAWYERS
IN LOS ANGELES

WINNING THE NOBEL Prize for PCR put me and my surfboard on the front page of nearly every newspaper in the world. By the time nightfall had swept once around the planet, a conservative estimate placed 328,716 small caged birds directly above my picture, flicking little greenish droplets rancid with uric acid.

That number is based on a world population in 1993 of 5,506,000,000 and estimates by Mr. Jamie Yorck, a noted bird expert from San Francisco, that one sixty-seventh of all human beings own caged small birds; and all of them of necessity buy some newspaper, which they replace daily in the cage, unless they are pigs, again according to Mr. Yorck. The average size of that newspaper might be twenty-five pages on weekdays, and my image spanned about one-tenth of a page. Thus, 328,716 birds were directly above me. Add the further indignity of chicks, kitties, puppies, wrapped fish entrails, and the unknown dark fluids deep in tropical dumpsters. It gave me the creeps.

Speaking of birds, I bet you didn't know that birds and primates, but not kitties, puppies, or other mammals, secrete uric acid as the final metabolic product of DNA. Uric acid secretion is something primates have uniquely in common with birds. Among the primates, humans are the only whistlers. Birds whistle. I think this suggests an avian connection, and isn't it true that we have learned to fly? Maybe again. How come we're not descended from the birds? Perhaps, if the hulking, upright, bipedal dodo had not been extinguished so quickly in the seventeenth century by the arrival on Mauritius of European fools with firearms, there may have been more scholarly debate on the missing link status of the dodo. DNA evidence as advanced by PCR in the latter part of this century made our avian ancestry less likely. On the contrary, it supported our very close connection to the apes. In 1995, in Los Angeles, California, our collective behavior in the O. J. Simpson trial was not helpful in refuting that connection. If we hadn't been already, then there we made apes of ourselves.

The Nobel Prize splattered me a bit, birds and all that. But if the O.J. trial did also, then there, at least, I was not alone.

Among the evidence found at the murder scene were several drops of blood not from the victims. Presumably from the killer. DNA tests conducted by the prosecution indicated they belonged to Mr. Simpson. It placed him at the murder scene. It was the most incriminating evidence against him. The defense lawyers knew that if they couldn't raise doubts about that, O.J. was in trouble.

Mr. Simpson hired a number of lawyers, including Robert Shapiro, Johnnie Cochran, and F. Lee Bailey. Because of the DNA evidence, they brought in Barry Scheck, Peter Neufeld, and Bob Blasier.

It didn't surprise me when I got a call from them. Barry and Peter wanted to come down to La Jolla.

In the American judicial system, you don't come into a trial as a neutral observer for the court. You have to be on one side or the other. You can't just be an expert. You have to be for somebody. It's called an advocacy system, and it's a little weird. You swear to tell "the truth, the whole truth, and nothing but the truth, so help you God."

"So, help me, God. Do I tell the truth differently depending on which side I'm on?"

"Sorry, pal, can't help you there. You want to talk about sin, I can do that. You want to talk about the law, get a lawyer, or talk to the other guy."

You work on it in the wee hours when you have to get up and testify. You get to know the system and you discover that the "whole truth and nothing but the truth," regardless of the poetic flow of the oath, is not what they are expecting. And they do have their reasons. You can especially forget about the "whole" business. Only selected parts of the truth are of interest. And only certain parts are required. What is required is determined by a dusty web of law and precedent stretching back to England, and then by the daily pleadings of the lawyers, and then by the learned rulings and whims of the judge, and always waiting just offstage, the sudden and unpredictable turn of the cards.

If you are a lawyer, you have started pondering these things in school, and as your career advances, they seem more and more reasonable. Advocacy is at the root of this. It is a very large and very loaded word. It is our system of justice. We don't trust our system and we need an advocate of our own. And if we can afford it, we need a damn good one. Maybe several. If he's been on Larry King, so much the better.

I had previously testified in murder trials for the defense, and I'd felt that my role there was to make sure the PCR-DNA work had been done fairly and correctly. I was not there to be on someone's side. I found in almost every case that the testing protocols did not stand up under careful scrutiny and that the errors were neither inconsequential nor insubstantial. Was I falling under the spell of advocacy? I don't know. I think I was being objective.

Technical testimony by an expert witness, ironically enough, isn't. It's very much a matter of style, not content. You can't talk to a jury about the technical details of your specialty and make any sense. The jury won't know what you're talking about, and that is precisely why you were hired.

So what you say is much less important than how you say it. It's like when you get off the beaten trail in Mexico—you can't speak Spanish and they can't speak English—but you don't start acting like an asshole if you want some help. You act helpless. And they help you. You make your points with your hands and your eyes, and the tone of your voice.

Cross-examination sometimes comes down to the opposing lawyer forcing you to give a simple answer to a complicated question that you would rather answer with a yes or a no followed by a long qualification and some mitigating circumstances. The cross-examiner will hold you against your will to a simple answer. He'll keep asking the question over and over. Your counsel will keep objecting. The jury sees you struggling. Sometimes the objections will be sustained, finally one of them will be overruled and you will have to answer. If you say "Yes, but" followed by a long explanation, chances are your explanation will be stricken from the record. It looks as if you're trying to squirm out of something.

I have mixed feelings about expert witnessing in a murder trial.

Barry and Peter came by La Jolla, and I liked them right away. They were not super-cool-plenty-slick lawyers in it exclusively for the money. They were law professors from New York and had been using DNA evidence to get innocent people off death row. They told me that the newly developed DNA tests introduced into reopened trials established the innocence of one out of four convicted but unconfessed murderers. One out of four? I was surprised.

If you find DNA in intimate association with a crime, for instance, in the underwear of a rape-murder victim, and it's not the victim's and it's not the suspect's, and there is no DNA present from the suspect, then you've very likely got the wrong man. On the other hand, just finding DNA at the scene of a crime that resembles a suspect's DNA in every way you have examined it could mean many things. If you find the first two numbers of a social security number you can prove it's not mine if it doesn't match, but you can't prove it is mine if it does. You need the whole thing to do that. DNA evidence as obtained by forensic labs is only the first two numbers. It has its limitations.

Like just about everyone else, I was curious about the Simpson case. Barry and Peter said the DNA work had been totally botched, and after they showed me some details of their analysis, I could see they had a case. Not only had things been botched, but the honesty of the detectives, Lang and Fuhrman for two, was not at all clear. That wasn't really my job.

I got involved and I looked closely at all the details.

The Los Angeles police didn't know how to run a lab and

probably shouldn't be expected to know any time soon. The very idea that a lab is run by one of the advocates in an adversarial contest is itself a little fishy. LAPD labs had some of the right tools but by no means all the right tools. People had been hired to follow the written instructions on the boxes of DNA investigation kits available from various manufacturers, but the people were young and they didn't know the chaff from the wheat, or their ass from a hole in the ground. They were fresh out of college.

After looking at facts, I decided most of the DNA evidence should be thrown out on first principles. I'm referring to principles of science that had been clearly established by the end of the seventeenth century. Nothing fancy. I agreed to demonstrate to a jury why I thought that.

Once I had said yes, I started watching the trial, and once I started watching, I got hooked. Me and the entire AARP and everybody who was unemployed at the time or only had to work at night. The mothers of everybody I dated after that loved me. I could talk about the case. Sadly, the daughter had to work. It was the most incredible soap opera ever. Did it get awards? It certainly spawned a bunch of lawyer shows.

I knew all of Marcia Clark's outfits. I noticed when she bought something new, and I was horrified when she changed her hairstyle. I figured that when I was finally up there I would probably have lunch with her, the way professionals do, and tell her, casually, just between us, that I thought the hair change was distressing for the jury's sense of continuity.

When I finally ran into her, we were walking in opposite directions on the aisle that separated the accused from the injured side of the courtroom. I smiled at her. I felt as if I knew her. She saw me. She wasn't familiar with my wardrobe, but she had already been talking about me on TV, and she had seen pictures. She knew me. And she knew that I had just arrived in town to work on her case. We were both being paid. I expected at least a weak smile.

She looked straight ahead, right past me. Precisely straight. Not a single degree off line of 180 degrees from the direction of her rear end. I was deflated. She was like that the whole time I was there. Day after day I watched this princess of the county unable to feel even the first degree of noblesse oblige. Oh well, it was Los Angeles. Darden was equally unapproachable. I think he must have modeled his behavior on hers.

When you get the hang of it, science, like everything else people do for a living, is pretty straightforward. You are in the business of solving puzzles. The way to approach a puzzle is to think about it for a while, look at all the facts you can find out about it, and then take a guess. Propose a solution. The next step is to try your best to disprove your solution. Show that the pieces don't fit together in the way that you have proposed. If you can do that, then propose another solution. And then do the same thing. Reality is a tricky little puzzle. Sometimes a few pieces will fit together but they don't really belong together. Some solutions will seem to be right for a time, but then they fail. The one that accounts for all the relevant facts and cannot be disproven—all the pieces fit together without squeezing them too hard, and new pieces fit together on top of them—is probably right. It's as close to being right as your ability to know the initial facts. You can claim that your solution is tentatively true awaiting further study. Or look at the picture on the box.

The crime lab is supposed to help figure out whether the pieces fit together. It isn't supposed to make sure that they do. When the experiments conducted in the crime lab come out black and white, it's easy, but often it doesn't happen that way. If the results of testing are not quite clear, forensic scientists have to rely on a large number of observations that are maybe yes—but quite possibly no—and then a new dimension comes into the process. Self-delusion.

It is here that people have to be very careful not to get personally involved in the act of deciding between alternative solutions to a puzzle. Their salary, for example, should not hang in the balance.

The way good scientists deal with the possibility that they may be swayed one way or the other in their evaluation of fuzzy observations is to intentionally keep themselves totally in the dark about the meaning of the observations they're making. This is known as doing a blind study. If, for instance, you want to sell a new drug and you work for a drug company, the FDA requires that you do a blind study. They don't trust you. And neither should you.

In the Simpson case it would have been prudent from the very beginning if O.J.'s blood had been stored locked away in a coded vial without his name on it. To make things convincing, several other samples of blood should have been taken from presumed innocent people, coded and similarly stored. The DNA structure of all of these samples should have been compared with that of the blood found at the crime scene. When all the testing was done, then and only then should we have opened the envelope.

The courtroom is hushed. The coded labels on the tubes are

revealed. The lawyers and the judge at the side bar open the envelope. If it had been agreed that the blood on the walkway came from the person whose blood was numbered LAPD004, and that just happened to be the tube containing the blood of Marcia Clark, everybody would have giggled and the DNA evidence would have been dismissed. If, on the other hand, tube number LAPD004 was Mr. Simpson's, the case for the prosecution would have been immeasurably strengthened. The way they did it was like a one-man line-up

This is the kind of thing I would have addressed had I been called to testify. Another thing that would have been reasonable for the LAPD to have done, and something that should always be done in DNA cases, is very simple and obvious. Just to keep things fair, blood samples should be taken in the presence of an advocate for the accused. Some easily measured and impossible-to-remove chemical—they are called tagants in the chemical business—should be added to the sample. It could be blue food coloring from the convenience store across the street, or better yet, it could be a DNA tracer made specifically for this purpose. Companies sell them for less than a hundred bucks. They are impossible to remove without removing the DNA. That way, if questions about the chain of custody of a DNA sample come up, the questions can easily be resolved. When it was suggested that Simpson had been framed, that blood found on the gate at the murder scene had actually come from a test tube of Simpson's blood that Inspector Lang "had kept in an envelope on the back seat of his car for several hours," a tagant would have made it easy to refute that. Instead, we were treated to interminable hours of testimony from both sides about the presence of a chemical called EDTA, which proved nothing and which nobody understood.

I saw O.J. only in the courtroom. They had him locked up at night. He was presumed innocent, I reckon, but considered to be a little dangerous. They might have made better provision for him, like guarding him in his own house, just in case he won. I guess they were still a little put out about him driving around on the Santa Monica Freeway with a pistol and his friend Al.

There was a line on the floor of the courtroom. O.J. was confined behind it by a congenial marshall who seemed to enjoy his job. He had a good seat. Only lawyers were allowed to cross the line. But when court would adjourn in the afternoon, the lawyers would crowd around O.J. and the horde would extend back over to the witness side of the line and the marshall wouldn't notice who was who. O.J. noticed me immediately. The first thing he asked me about was my refrigerator.

An article about me in *Esquire* had included a photograph of my kitchen. On my refrigerator were pictures I had taken of several women who had passed through my life, some of them without clothes. O.J. asked me about the woman who graced the upper right-hand corner. "She's gotten married," I told him. He offered his condolences. I never asked him about the killings. It's a tough subject to broach even in a courtroom.

I passed him a note one time through Peter. There was some former cheerleader appearing as a material witness who had a tape of a phone message O.J. had left for her on the day of the murders. As was their habit, the lawyers had buzzed around it furiously like flies in the backyard of an untidy dog, and Ito had finally decided to admit some of the text of it. The cheerleader had appeared in court that day looking somewhat bewildered. She was a breath of fresh air, and I helped her get situated. While she was testifying, I scribbled a note to O.J.

inquiring about her phone number. He replied, through Peter, that in order to maintain my loyalty to getting him out, he would rather not furnish it. He would be interested himself, once this was over. Throughout the trial, O.J. maintained this level of playful charm and humanity. It just wouldn't mesh with the awful things that occupied us there.

Almost from the beginning, even after I arrived in L.A. and started working with the lawyers and the other expert witnesses, there was a question of whether I would be called. Rockne Harmon, a deputy D.A. from northern California, had made a career out of handling the DNA evidence for prosecutors in various jurisdictions around the state. Rocky had a reputation for playing dirty pool with DNA witnesses on the other side. Most of it he probably deserved. When he heard I might testify, he made a clumsy pre-emptive strike on Court TV in an area that he must have assumed would make me personally uncomfortable. It concerned my past LSD use. His attempts certainly made my mother uncomfortable. I had made her a little uncomfortable myself, talking about LSD in a magazine interview the year before. Rocky had picked it up from there and decided it belonged on the evening news. That was right in line with Mr. Harmon's well-known method of responsibly prosecuting a case in the name of the People.

Neither my mother nor Rocky understood that it didn't make me feel uncomfortable. I pictured myself the honest self-admitted defender of LSD use and clearly on the high moral ground. If I took my honesty seriously enough to admit that I had broken the law by taking LSD, why would I lie about anything else? I was pretty much alone in that viewpoint.

On the phone from South Carolina, Mother pleaded. "Now, Kary, you don't need to be telling them about that, do you?"

"You want me to tell the truth, don't you?"

"I certainly do. Tell the truth but not that truth."

I told her I had no choice in the matter. Having a fairly poor memory for certain details, I had discovered early in life that the truth is much easier to tell than anything else. It saves a lot of confusion. "I have been a difficult son in some ways, Mother. But don't you love my simplicity?"

"No."

"Sorry."

Theoretically the jury did not know that Rocky had suggested in the courtroom that he would like to have some assurances that Dr. Mullis would not be on LSD while testifying. That kind of foreplay was allowed only during their absence, but everybody in the world got to hear it replayed over and over, and sequestering a jury while allowing conjugal visits from spouses is a little preposterous. Do we really believe that the twelve people most curious about what's going on know less about a case than a literate canary could glean from the morning paper?

Legally speaking, juries don't volunteer. They shouldn't be tortured. But isn't what we do to them now for the sake of our cognitive dissonance in regard to their supposed sequestration torture? Isn't it torture to subject someone "on his honor" and with "fear of retribution and public embarrassment" to conditions guaranteed to lead to either their non-compliance or the break-up of their marriages? As a public concerned about fairness, we should put a little effort on the "Plight of a Juror in the Age of Long Highly Publicized Trials." At least we shouldn't be snotty when we catch some juror failing in his public duty.

We have another problem—jury selection. Are you supposed to be tried by a jury of your peers—or by a bunch of people that a professional jury selection team chooses from hundreds of candidates based on the probability that they will vote one way or the other? What's going on here in the name of professional jurisprudence?

When I first appeared in the courtroom a week or two before I was scheduled to testify, I looked at the jury. I had never seen them before because the TV camera had been intentionally located above them, but they knew me. Their eyes said, "Boy, are we gonna have fun with you!" I nodded agreement. I, too, was looking forward to injecting a little candor into the drama. Most people are grateful when you educate them about anything other than their own flaws, and I figured we were going to do that. Because of the Nobel Prize, I am a free agent. I don't owe anything specifically to anybody. I think it makes me a good educator and a good witness.

I also know how to speak to an audience. I don't think Rocky Harmon realized that. It didn't say, in their collection of documents describing me, that I had a rare talent for cutting through the garbage to the issue, to pull the forest out of the trees and into the light where the jury could see it for what it was. It didn't say that I could help them understand, without their knowing anything about chemistry and blue spots and statistical analysis, just what had gone wrong in the LAPD's lab. I think he figured I was the flake he claimed I was, and he was not afraid of me. He wanted me on the stand as much as I wanted to be there, and we were both disappointed when it didn't happen.

I had had a plan of my own for Rocky Harmon on the stand. I had explained to Barry, after reading about Rocky running roughshod over defense witnesses in previous cases, that I thought it would serve him right if we allowed for him, who had lived by it, to "die by the dirty sword." Barry has a sense of humor and, for a lawyer, an amazing sense of justice.

I was going to let Mr. Harmon decide, before the courtroom, that what one had done in the past did indeed have relevance for judging the truth of what one had to say in this case. I would do this with a question that would be out of order, because lawyers are supposed to ask the questions. If I was lucky, he would think I was trying to avoid his questions about my life, and he would quickly answer that yes, it was relevant. Then, with the Q&A rhythm on my side, I was going to slip in another question of my own—about some outrageous thing he had "done" in the past. Something like an incident with two young boys in the park. I would be extremely out of order—he would loudly protest—and Ito would slam his gavel down to silence me. But the answer wouldn't matter. The question, like so many Rocky had asked in his days of fervent prosecution, would be transformed by the alchemy of the courtroom into a statement.

Whatever happened after that—maybe a fine for me, maybe a movie contract, or maybe a night in jail—I think my points from direct testimony would have survived intact. Pressing my luck further, I would conjecture that the jury and most of the courtroom would have been happy to see Rocky catch some grief in exchange for what I considered abominable misbehavior. Any further interaction between Mr. Harmon and me would have been related to the DNA issue only on the surface. Underneath it, there would be those two boys. He had been trying to set up a diversion, centered around my LSD use. I think LSD would have paled beside fictitious young boys. Innuendo does have its charm.

Sadly, my favorite cross-examination never happened.

I was dropped from the roster at the last minute. For those few who were concerned about me—yes, I did get paid. The amount is a professional secret, but I drove back to La Jolla in the same 1989 Acura Integra I had come in.

Johnnie thought the jury was convinced and saturated on the DNA issue before I was scheduled to testify. He could probably tell by the way their eyes glazed over when they heard the DNA discussed. They had been subjected to tedious and technical testimony for weeks. The prosecution had certainly failed to fulfill its obligation of proving the value of the DNA evidence beyond a reasonable doubt. I felt that the jury would be even more convinced after hearing my testimony, and I thought some of the things I would say could influence the whole case. But the question Cochran rightly asked was, do we need to take a chance by going further when it looks like we've already won on DNA?

My horoscope says I shouldn't expect to be a corporate kind of guy, and I'm not. Bob Shapiro and Johnnie Cochran couldn't help liking me—I have good manners, I'm well informed and often funny. But there was a danger there, and they didn't get to be successful lawyers by holding their hands over their faces and saying, "Well, here goes."

In the light of day, the scientists had had their moments and the issue of the physical evidence had settled somewhere between "indeterminate" and "contrived if I heard it right." It was a long way from the gory trail of blood with which Deputy D.A. Clark had opened her case.

Things were looking good for O.J. The prosecution was starting to fade. Marcia's hair needed upgrading. But where can you go from curls?

I packed up my white shirts and coats and ties and rolled them down to my Acura, where the valet parking people probably didn't know that I wasn't going to testify. On the way out of town I stopped by a strip joint that I had frequented years ago when I was working in L.A. In the last few weeks, I had become aware that people recognized me. I decided I wouldn't let that bother me.

I was having fun in the club, losing myself in the pleasures of the flesh. Every one of them was different, every one of them a little story in herself. I am easy to entertain, I thought. The possibilities are endless—tropical trees, sunsets, waves, females, quantum physics, crimes, biochemistry—then a flash-bulb drew me out of my reverie. A woman sitting across from me had just snagged a photo of me enchanted by a pretty girl dancing in the nude. The bouncer was on the photographer right away. Cameras were not permitted. I thought he would usher her out. He took her camera and brought it to me. "What should I do with it, Doc?"

I was now known as the "DNA Doctor."

After a few months, I would be less and less surprised. That night at the club I decided that my new public identity didn't matter, and I wasn't going to start acting any differently. "Give it back to her," I said to the bouncer. He delighted in the concept, and he took it back to her without pulling out the film. It's probably the only time he ever allowed anyone to get out of there with a photo. I nodded my head to the woman with the camera and continued enjoying the show.

I was ready to go, but I stayed another half hour just to let the bouncer know that I wasn't leaving on account of the picture. When I left, he followed me out to the sidewalk. We chatted a few minutes about the case and whether I thought Simpson was guilty. I agreed with him that it would be good to hear O.J. testify. But I had learned a little bit about justice and police and people accused of crimes. We didn't really have the right to make him come upfront and talk to us about it. Furthermore, we don't have the right to conclude that, because he didn't want to testify, he was guilty of something. At first it seems like a strange law. The defendant, who must know his whereabouts that night, is not required to tell us.

It seems weird until you think about how horribly askew the scales of justice can be. It's a balance. In O.J.'s case, he was able, by spending most of his money, to bring a defense together that had more clout than the state, and more persistence. Persistence is the state's big card. It's what bureaucracy inevitably substitutes for brains. O.J. bought brains and persistence. Most of us couldn't afford it.

Back on San Diego Freeway, heading for home. I figured that most trial watchers would remember me as the guy who smiled directly into the camera and waved. They were talking about me that day in the courtroom. I was sitting in the audience, and the camera had found me. I could see where it was pointing. Why should I ignore it? I looked up to the TV watchers of the world and waved. My mother thought it was cute.

Another day Mother had called and warned me against sleeping in the courtroom. "You were sleeping this afternoon," she said. "I saw you."

"Mother, what the hell are you talking about?"

"You had your head down on the table," she continued. "They had the camera on you. It looks disrespectful, Kary, and that man was talking about DNA."

"Oh. Shit!" I realized what she was talking about. "Mother, there's a TV monitor right under the table in front of me that lets me see what's on the screen in front of the jury. I have to lean down to see it."

"Oh." She laughed reluctantly, without losing the offensive, "But it looks like you're sleeping."

My mother is seldom impressed by alternative explanations of her own discoveries. I kept her observation in mind for future reference. She also doesn't approve of my language, although I explain to her that it is expressive.

Back in San Diego, I thought about what my least favorite image from the trial was. When the verdict was announced and the tired jury was released, the district attorney of Los Angeles County, who had mostly stayed clear of the courtroom, roped all the prosecutors together, and in front of the cameras, they cried. They cried because they had not been able to convict O. J. Simpson.

Did it ever occur to them that the reason we had a trial, the reason we hired them in the first place, is because sometimes the man that the police arrest is innocent? Is it reasonable for a prosecutor to cry when a long-suffering jury comes to a decision and acquits a defendant, who is therefore by our highest standard innocent?

I just hope that I don't ever get arrested back there.

In La Jolla I live on the beach across from one of the nicest surf breaks in California. Waves are intricate things. Our best waves in La Jolla are born way out in the Pacific when a storm, maybe as far away as New Zealand, a storm with a low pressure area, pulls up a huge convex mass of water from the ocean. It literally sucks it up over a period of days, maybe a few feet above sea level at the center, and the water falls back into the sea, and when it falls it pushes up more water around it in a circle, and that process of water falling and rising in expanding rings travels away in all directions, like the circles of waves that rocks falling leave in a pond; only it starts with suction near New Zealand. One of the directions it travels is La Jolla. And the winds are often blowing this way, too. They blow on that rising and settling expanding mass for a few days, and the friction of the wind over the disturbed water brings up lines and lines of nice waves.

We always surf in the morning because the wind is quiet. The waves are glassy. We paddle out. We sit for a while and talk. Then I see one coming. The peak is aimed just toward me. I paddle hard to get the board moving. Steve, my friend, offers, "This one's yours, Mullis."

Maryjane agrees. "Go for it."

I paddle. It seems sluggish. It's been a while. The wave lifts me up, I dig into the water, and with a last stroke, I pull myself over the top and plant my feet on the board. I'm there. The wave is holding me up and at the same time rising up behind me. I'm in control. I can cut to the right or the left as if I were skiing down a long hill. I don't want to go straight down the front of it. It could come up behind and pitch me over. It's much more fun to take the sharpest angle I can muster, like a sailboat digging into the wind, feeling the power of the angle and screaming across the face.

When I fall, I hold my breath. The sea takes me into her

arms. I'm not bruised. I wait underwater until the action of my board in the whitewater, bouncing around, looking to all the world like it was trying to find my head, is over. Then, I come up.

My friends laugh. They think I'm a bit rusty.

I've been away. It's good to be home.

5

THE REALM
OF THE SENSES

M OST OF US agree that we have five senses—five
tiny windows—and we are locked in our own huge
castle looking out through these five tiny windows.
The world outside, we gather, has no limits except this one little one—it has no end to its hugeness or to the minuteness
of its details, or to its tangled vines of complexity that coil
around themselves from forever in ways that only very young
fools would conspire to untangle.

One of our windows is hearing. When the air around our
ears goes back and forth 50 or 60 times a second, it causes our
eardrum to sway gently back and forth, and we hear a very
deep hum, like the sound of AC power out of the wall outlet in
America, getting into your audio system. If the air pulses in
and out around 880 times per second, the eardrum vibrates
and we hear a sound like the middle A on a piano. At 2 times
880 equal to 1760 cycles per second, we hear an A one octave
higher. At 3520 cps, we hear the next A up the keyboard, and
so forth. At 20,000 we no longer hear sound, even though the
air around our ears is still vibrating. Some component in our
detection system fails. Although air can vibrate at even much
higher frequencies, we can't hear it. Dogs can hear higher

pitches than we can, children higher than adults, and women higher than men in general. Our window on sound is narrow. It is pretty much centered on the range of sounds that our bodies can make. Our ears must have evolved mostly for the purpose of listening to ourselves.

Our biggest window is vision. In this case we are tuning in to the vibrations not of air but of something we call electromagnetic fields. If you could wiggle a little magnet 428 trillion times per second, it would start making red light—not because it was getting hot but because the magnetic field was oscillating back and forth. The magnet could be cold. I don't know anybody who could actually wiggle a magnet that fast—this is what is called a thought experiment. The point is that the magnet is not getting hot from the friction of the movement in air; it could be happening in a vacuum. It is the back and forth motion of the magnetic field around the magnet that is making the light. A little faster, 550 trillion times per second, it would glow green. At around 800 trillion times per second, even in a dark room, the light from the wiggling magnet would no longer be visible to you. A little faster, and it might start causing sunburn on your face, but you could no longer see it. It would be what we call ultraviolet. Any time a magnet wiggles, no matter how small it is, or how fast or slow it wiggles, it makes some kind of light. Most light is made by little magnets called molecules, and our eyes are tuned to a very narrow range of it. Our vision is centered on the 550 stuff we call green, because we developed our vision while we were living under the canopies of giant trees that let in only the green light.

We also have taste, touch, and smell. From birth, we also

have the ability to detect "weightless." We don't like "weight-less." Unless we are in orbit, it means we are falling and going to land soon, maybe hard. If we are in orbit we are still falling, but we are moving so fast that by the time we fall to the level of the earth, the earth is behind us and we miss it and just keep on falling. We go around and around like the moon, which is also falling and sliding past at the same time. Our sense of "weightlessness" is not one of our more pleasant ones. It doesn't have a lot of comforting familiarity of detail about it. Either you are or you aren't falling, so it's not much of a sensory mode by itself.

It does, however, accentuate one more frequently acknowledged sensory mode. It accentuates waiting—that is, the sense of time passing. We can count seconds in our head. The notion of waiting and marking time becomes severely hyperactivated by the sense of falling.

And those definite five, plus the dubious two, make up the whole of our acknowledged sensory modes. All of our windows out onto the vastness of outside, from the vantage point of our castle—prison or hermitage, depending on your personal bent—are described in our various languages in terms of those seven modes of perception.

Our brain is accustomed to listening to the news from the windows. Our thoughts are at their clearest when the input is from our eyes, then ears, then nose—maybe nose before ears—then tongue, then skin—maybe skin before tongue. It depends on what you're licking.

Some species have other, somewhat amazing to us, sensory abilities. They are amazing because we have this inaccurate perception that everything that is real is perceptible by at least one of our senses, and invisible things are kind of freaky. A bumblebee can find its way, supposedly, by observing the polarization angle of light, which we ourselves can't observe without equipment. Dogs can sense when their masters may be in jeopardy. Mailmen are quite familiar with the phenomenon, which, in their case, is dysfunctionally aroused in the dog unless they are bringing a notice of foreclosure. Dolphins use sound to visualize three-dimensional space. Sharks, I am told, can smell a drop of blood in water hundreds of feet away. Ants sense one another's needs well enough to work in huge teams. Bats navigate by sonar.

How do I know when someone is standing behind me even though I can't see them? I can't hear or smell the person, but I have a sense about it. I also have a non-visual, non-scent-related, non-intelligence-linked sense for finding my way out of the woods at night, which is convenient because I'm in there a lot and it can be very dark in Mendocino when it's foggy.

Intuition is used to describe those odd feelings we get from time to time that cannot be ascribed to our five favorite channels. We don't have names for the remainder of our senses, and they have gotten a bit of a tawdry reputation because of the amazing success of the five that do have names. We have made a lot of cool things, some of them charming and some of them horrible, using the logic that developed out of those Fantastic Five. But perhaps the most important thing that those five senses and the rules of mathematics that we created from them have told us is that they cannot tell us everything. They are a narrow slit, swept only briefly through a glass darkly.

What do senses have to do with logic and mathematics?

In the mid-nineteenth century, anybody who thought about

it thought science was based on mathematics, and mathematics was an abstraction based on sensory input. It was assumed that mathematical laws, and in particular geometry and calculus, were at the root of the way things worked. The parabola that Newtonian physics predicted for the trajectory of a cannonball could be trusted to drop the cannonball on the heads of the Prussians if necessary. The relationship between the charge in the cannon, the weight of the ball, and the angle of the shot would faithfully define the parabola. Only a few people have the slightest idea how to deal with math. Most of us just light the fuse.

You probably didn't like math in high school and you probably still don't. You also probably think math is arithmetic. Arithmetic is the part of math that is useful for balancing your checkbook. Most professional mathematicians can't balance their checkbooks. Some of them are involved with trying to understand whether a basketball could be turned inside out—if the only restraint on the substance of the basketball was that it could not be crinked excessively. It could, for instance, be pushed through itself. It just couldn't be crinked—without causing any folds in it from becoming infinitesimally thin. Remember that this is abstraction supposedly based on sensory information. Sounds bizarre, doesn't it? The very few strange people in the world working on this are called mathematicians and would be further classified as differential topologists. What they are doing is similar to what most theoretical mathematicians are doing. They are taking the things that our senses can tell us, drawing some conclusions from that, and then trying to understand things that are not, and may never be, discernible through our senses. It may have something to do with whether we can make a device that will transport us to Venus in a heartbeat without stopping our heart or, more to the point just now, make our computer keyboards stop freezing up. We all know the aggravation of having none of the buttons or devices on the computer function except for the plug—the final act—and then when it comes back on, it has the nerve to suggest that you turned it off wrong. In some way, unknown to you or me, the guys thinking about inverting basketballs are really thinking about this. It's called mathematics. It derives from our senses. Maybe. Things have a tendency to drift away from their origin.

Pay attention to your senses. Neither differential topology, nor geometry, nor calculus has turned out to be the real underlying root of how things work. The rules of geometry and calculus were derived from sense perceptions and can be applied to the things that usually concern us—throwing a baseball, shooting a missile—things that we can access and confirm with our five senses, but they are not the reality that we consider in this century to underlie our lives. And this is an important sociological point. At a certain level in physics— the realm of the smallest things—calculus means nothing. It is too dependent on time and space. Time and space don't really count for much in the inferno of the very small things that we now think are fundamental. So senses don't either. Geometry doesn't work at all on the really small things.

Maybe this is saying something important to us. If geometry doesn't work, then our attempt to understand things way down there in weirdness space may be inappropriate. Sensory perceptions have nothing to do with things there. Sensory perceptions confirm calculus, and calculus doesn't work there either.

We can't do any engineering down there. Nobody who is sane understands what goes on down at the level where the fundamental things like quarks and electrons do not have any volume or any position. If you can understand something with zero volume and no position, then welcome to insanity.

Do we need to use billions of dollars to build machines that maybe will put a few of our rightfully treasured eggheads in touch with things so far from what can be engineered into useful items that only they will get a thrill out of finding them? Do we need to do this when there is an obvious threat over our heads, something falling right now onto our planet? Something big, heavy, and headed our way. Something that already has our number on it, a number we could read if we would just point enough telescopes out there to see it.

I once heard, and I think it is true, that only one man in the world—some Indian mathematician—understood the mathematics of string theory in eleven-dimensional space, and he dreamed it. That may be an exaggeration, but it isn't far from the facts as I know them. We would need a big machine to find out whether he was right.

We humans, including mathematicians, have an idea. It is that the smaller something is, the more fundamentally important it is. And the bigger something is, the more fundamentally important it is. Maybe we want to reconsider what we mean by fundamental.

There is an important story here. It is the story about how, as a culture worldwide after the big wars, we have begun to drift into the idea that reality is not what you see with your senses. That reality can be seen only by specialists with heavy lenses and special machines. It doesn't seem to matter that for millions of years we've been developing some of the best sensory apparatus in the solar system. It grows in the wall of the castle that forms around us as soon as we are conceived, and unless something fucks it up, it works quite well for fifty or more years.

When did we as a culture decide that extremely little things were fundamental? I think it was this century and the advent of nuclear bombs. At the same time, we decided that very big things were also important. Medium-sized things like us were relegated to the not-so-important closet. How did that come about? Academic departments like Aesthetics and Existential Philosophy vanished without a trace. Their questions about medium-sized things were still largely unanswered. Medium-sized things are still pushing grocery carts around full of their last possessions, international diplomacy still involves threats of explosions, and nobody knows what the weather will be like next fall in Florida.

We are out here, culturally, somewhat alone today. We have no counsel from colleagues writing from ages past about this kind of thing. It is a place where we have arrived, hell bent on knowing what's going on everywhere but inside of us, or even close by. We have applied geometry and calculus over and over. We have built machines that can see farther and deeper, and we have analyzed the results we get from these machines farther and deeper, by building computing machines. And now, in subatomic nature, we have found a set of structures that don't look like anything we know. They don't look like cubes or spheres, or tetrahedrons, or even sexy little dodecahedrons, or horrible icosehedrons. They don't even look like zebras. They don't look at all. They come in and out of existence unannounced—no RSVP, no clue as to why they are there or why they wouldn't be there.

Also out on the cutting edge of physics are the cosmologists—the physicists who thrive only on the things that can be described as 1 quadrillion times the size of a basketball. It's a different part of the cutting edge. It's cosmology, the part of physics described as "examination of and speculations concerning the whole universe." It is explored by viewing light that was emitted billions of years ago, which we couldn't see even if we were on some mountain where a big array of detectors was looking at it, because that kind of light doesn't even register in our eyes. We can think about what it means, but we can't see it. It's exciting.

But the comets are falling. They are on their way right now. We can't see them in the moonlight yet, but we can predict them. We are a smaller planet than Jupiter, but we are a planet just the same, and things have often fallen on us. Some of them have been rather large. One of them, 65 million years ago landed off the coast of Baja and sent a tidal wave five hundred feet high over Kansas City. The dust from that impact resulted in a hundred years of darkness on Earth that caused 99 percent of all species to vanish. And how often do gigantic things like this happen? How often do things like the crash of comet Shoemaker-Levy 9 in 1994 on Jupiter happen on Earth? What causes all the craters on the moon? Do we think that craters don't happen here just because the evidence washes away in the rain?

Isn't it reasonable that we should put some of our brightest minds on this? Won't it seem short-sighted if we look up at the sky one day and say, "Holy shit, we're done for now!" Three comets will be falling on us tomorrow, and there isn't a damn thing we can do about it.

About 6 billion people will die in one or two minutes following the collision. We could have been ready with any number of technological solutions—mainly missiles with big bombs—but we had failed to notice that our existence here had never been guaranteed. We weren't ready. We had been distracted by the pleasure of being the kings of existence and the inventors of thought.

In 1992 the Earth-orbit-crossing asteroid called 4179 Toutatis passed within 3 million miles of earth. It is 2.9 miles wide and it is coming back in the year 2004. They watched it come by last time with the Deep Space Network Goldstone radar antenna in California and the Arecibo telescope in Puerto Rico. It may miss us this time by about a million miles—that's four times the distance to the moon—plenty of room. But it will definitely be back again, and it's hard to know what the precise orbit of an asteroid might turn out to be until it's really close and you happen to be in the cross-hairs. Maybe it would be a good idea if we paid a little more attention.

There are two really important things we could be doing now with our physicists. One of them would be to put physicists on the problem of things that may fall on our heads from space. How to detect them coming in and how to deflect them. The other would be putting another bunch of them seriously into action trying to get in touch with other cultures in space that have dealt with these problems and can help us solve them. Nobody has any idea about whether there are beings in space that are broadcasting, on seven hundred channels unknown to us, useful hints about how to take care of potentially destructive comets. The former solution is certainly something we can do and it doesn't rely on the unknowable existence of other cultures. Give it first priority—our boys can solve that one for sure. The latter solution, asking for help from someone who may not exist, is worth some effort, but it isn't a certainty. I'd put 90 percent of our present expenditures for physics and space technology on the former problem. It is crucial and we can solve it. The other 10 percent I would distribute among the latter problem and the trivial ones that involve our insatiable need to think that we can understand the fundamental nature of the universe.

I am not at all suggesting that we abandon physics. I am gingerly suggesting that we may be looking for it in the wrong place. I think we should fund the hell out of it. It has too many interesting twists and turns. I'm saying, however, that we should fund the kind of physics that deals with things that we can see and still don't understand, but once we understand, we can do something about. Why look any deeper? We can easily measure a lot of things that are puzzling. Weather is a good example of something that we should know more about. I'm not talking about global climate change. That's political. I'm talking about regular weather that has to do with whether there will be high winds in the Sierras next week. Very bright people should be directed into things that matter to them and to us—and that can be solved—rather than wasting them in fields that are very romantic but have very little relation to our lives and no relationship to the great referee of all things—our senses.

Don't lose track of the fact that 65 million years ago an

asteroid that had been pitched out of its orbit between Mars and Jupiter by a chance interaction with several other asteroids was deflected on a course for Earth that nobody at the time could even contemplate. In a few months it devastated Earth. Another lovely ecosystem almost completely down the drain. Oh well. Some lucky creatures always survive.

SOME PEOPLE IN Denver may survive the next Big One. They may have to eat frozen carcasses for a hundred years, fumbling around in the frigid dark while the fine black dust settles. Finally in summers, they may begin to see an orange area in the sky that in a few years may become the Sun. They may discover how to write again—first on the black in their caves, then on stone tablets. Then they may move south looking for salad, and a few thousand years after that, when they are once again on the verge of understanding the secrets of the quark, some ominous dark object may cross the full of the moon one night. In the morning, it is much larger, a star that should not be there.

We are no longer the tree shrews we were 65 million years ago. There are no more dinosaurs roaming Earth. But we are just as helpless against an asteroid as the tree shrews. The next time it happens, we will have to deal with something that the tree shrews didn't. We will have to deal with NASA showing us increasingly detailed pictures of our approaching doom and CNN speculating on how the stock market will react during the final few days. We will have to cry on each other's shoulders about our tragic simplicity. Do we ever buy an umbrella on a sunny day? Do we understand that our big blue and green planet attracts big mountains of rock from outer space with the same force that it employs to drive apples to the ground?

I say that everything, bar nothing, being a possibility, we are placing our bets right now on the wrong tables, exploring regions of reality where our senses are not reliably sensitive, when there are obvious things needing our attention right now, where our senses are reliable. Let's look out for long-term comets and straying asteroids. Let's go to Mars. Let's do bio-chemistry. Let's think about what kinds of senses people have and how they can use them to their advantage. Let's not spend our time and resources thinking about things that are so little or so large that all they really do for us is puff us up and make us feel like gods. We are mammals, and we have not exhausted the annoying little problems of being mammals.

6

I THINK,
THEREFORE I WIRE

I STARTED PLAYING with electricity when I was six. I was learning how to spell at about the same time. By the time I learned how to spell "volts D.C.," I had given up batteries for 110 volts A.C., which came out of the wall and never ran down. It made bigger sparks, but sometimes it melted the insulation off my wires before the circuit breaker tripped. The box was beside the back door in the kitchen, and it would shut down my plug and my mother's room at the same time. By day it was no big deal. I knew which switch to reset, and the circuit breaker kept no records. At night it was a different matter, but I had a little flashlight to find my way downstairs. From the dark of her room, my mother did not openly encourage me.

But while she said no, in fact, she placed no real restrictions on me. She made sure that each of her boys had a little closet of his own where he could do whatever he wanted. A storage room in the house that was not filled with useless things was assigned to each of us. Mine contained the water heater and a light bulb, to which I had added an outlet. I installed a magnetic lock on the door. To open it I would pass a magnet over a particular spot on the door. A nail inside would raise up and out of the eye on the end of a bolt that held the door closed. There was a string holding the nail so it wouldn't fall down into spaces between the roof and the insulation. No one but me knew how to do it and I didn't let anyone watch. I had a totally private place where I could launch my life when I was six.

I knew, maybe from birth, where the circuit breakers were. I don't remember Mom ever telling me not to do something specific with electricity, but she wouldn't have known what to tell me not to do anyway. From my closet I shut down the power several times. If the power suddenly went out in the house and my closet smelled like an electrical fire, she would tell me to stop doing whatever it was that I was doing, but she was not very specific, and there was a note of "please" in her voice.

When her Maytag broke, I took it apart. After I won the Nobel Prize she enjoyed telling reporters that as a child with an "overactive brain" I had ruined her washing machine by taking it apart. Mother knew a good story when she told it. The washer had been discarded and was sitting in the garage before I ever touched it.

I wanted to know what made the different cycles turn on and off. How did this machine know when to rinse? There was no power in the garage, so I couldn't do any experiments, but by looking at it carefully, I learned that there were little things along the water lines that I later learned to call solenoids. I learned this from looking at the wires and pipes. A solenoid is an electromagnet that can control a valve. Something that blocked a tube got pulled up when an electromagnet came on. The magnet worked only when the power was on, and that happened because a switch was closed. The switch in the washing machine was part of a notched wheel that rolled around constantly when the main power was on. It was a timer and I later learned to call the little notches castellations and the whole device a cam. It was cut out of a plastic disk that I think was made out of Bakelite, named after L. H. Baekeland, who invented plastic. Years later I heard that in the mornings he would stand out in his yard in a ritzy Florida neighborhood completely naked. The neighbors objected, but what do you say to the father of the Age of Plastic about his personal dress code in his own yard—in Florida?

When a stationary copper lever dropped into the notches of the Bakelite wheel, connections were made and power flowed through wires into the solenoids. It was very simple.

I took one of the little electromagnets and attached it to the cellar door. When I pushed a button I had wired up on the windowsill, the electromagnet would pull its little cylindrical piece of metal up and out of the way of a metal loop that I had attached to the door, and the door would swing open. The door would not have swung open, on being released, if I had not repositioned the hinges. I could let my dog out from under the house by pushing a button, and then, through the picture window in the den, I'd watch him bolt. I never knew why we had to shut him up under the house every night in the first place. I guess that's why I've always been an engineer and never an administrator.

B Y THE LATE 1970s my second wife had left and I was still living in Kansas City. I had moved there so she could attend medical school, but she ended up leaving me. It was a well-documented phenomenon and it didn't hurt my feelings. In fact, it was about three months before I realized that she had left. I was working in a laboratory at the University of Kansas Medical Center. I had a nice little house, and after she moved out, I filled the space she had occupied with electronic equipment.

I had an abundance of really cool stuff. Every month one of the labs at the university would close down—the funding had run out or a key researcher was leaving—and almost always the equipment was available to anyone who wanted it. I would go into those labs and salvage. Much of the equipment was old, from the early 1960s, and really wasn't of any use to anyone in the modern world. I could put it in the trash and then come back and get it, take it home, and there it was of some very serious use.

I ended up with a room full of equipment that would have been prohibitively expensive if I'd had to buy it. Things like ten-turn helipots, precision resistors, and power supplies that would provide precisely the voltage you dialed in with no spikes or fluctuations.

While at Berkeley, I had heard stories of people who could control their heart rate with their mind. They could speed it up or slow it down by thinking about it. People in India could put themselves into a state of hibernation. It sort of made sense to me. I knew that frogs were capable of burrowing into a hole and shutting things down for the dry summer months. I thought, I'm an animal—I could probably do something like that if I practiced it long enough. I decided to use my newly liberated electronic equipment to learn how to control some physiological parameter. I decided on the electrical conductivity of my skin.

The flow of electricity through a circuit is directly proportional to the voltage and indirectly proportional to the resistance. I put one electrode on each wrist and attached them to 9 volts. The resistance in my skin would range from about 14,000 ohms up to about 100,000. It would sometimes go higher or lower, but I was not able to control it at first. The salty fluids of my body were easy for electricity to flow through. The difficult part was my skin. I figured that the resistance between the two electrodes was mainly the resistance of my skin times two.

I decided to learn how to control the resistance of my skin. By just fooling around, I discovered that if I wanted my resistance to go higher, I had to think about something really boring or meditative. If I closed my eyes and thought about floating on a dark featureless lake, my skin resistance would go up. The voltage meter reading occasionally went as high as 180,000 ohms. But if I looked at a picture of a naked woman in *Playboy*, whoosh, the reading dropped below 10,000.

It was fun. To make it more of a pleasure to watch, I added an oscilloscope. I built it from something called a Heathkit. I bought a voltage-controlled oscillator at Radio Shack and put it in a circuit that contained my skin resistance. I could plot the variable frequency I obtained from the voltage across my skin against the 60-cycle frequency coming out of the electrical socket in the house and get some pretty interesting patterns on the oscilloscope. They looked like science fiction stuff and they were responsive to my thoughts. Bizarre. Wild tumbling shapes like the kind that happen just before the lab blows up. The technical term is "Lissajous patterns." If I really concentrated, I could adjust my skin conductivity so precisely that I could make the tumbling patterns on the screen stand still. That took a lot of practice, but it was definitely impressive.

Nobody else was impressed, however. It was hard to explain to a nursing student why it was amazing that I could cause a tumbling Lissajous pattern to stand still on the screen.

I realized I could do a lot of things with this. I could control a voltage with my mind. I decided to make a system that would allow me to turn a lightbulb on and off across the street in my neighbor's house. Technomagic. I hoped this would be impressive to a nursing student.

There were radio-controlled cars available in the hobby stores that had little FM transmitters in them that could send a signal across the street. If I used that transmitter to send a signal generated by my skin conductivity to a receiver connected to a set of transistors and a device that could drive a 120-volt lamp, I could turn that lamp on and off from across the street. I set it up with a circuit that would toggle the lamp any time my skin resistance changed rapidly because it was easier to drop it fast by looking at a nude picture of a woman than it was to raise it by not looking at one. It worked the first time.

People came by to see me do my telepathy. They knew I wasn't psychic, so they decided I must be an electronic genius. I decided not to argue.

The nursing students were impressed.

7

MY EVENING
WITH HARRY

IN 1978 I was working at the University of California–San Francisco on endorphins, which had recently been discovered in humans amid a good deal of excitement, because their natural existence in our body explained why opiates, like morphine, which had been found only in plants heretofore, could exert such profound effects on people.

It was about midnight when Harry showed up at my lab. I had been trying since early morning to purify an endorphin that wasn't very stable. Sometimes things you are purifying that seem to be unstable suddenly stop disappearing at some point in the purification. That means you have separated them from something else that has been actively destroying them. You hope this might occur, and you work rapidly for a long time trying to arrive at that point. I had got there around eleven thirty. I had my endorphin in the freezer and I was shutting things down. I was proud of myself because I had introduced a really clever step in the purification that had worked extremely well. It had to do with the fact that tetrahydrofuran is completely miscible with water until you add salt. I was eager to tell someone about it who would understand. Cynthia, my wife, would have believed me when I told her I had done something clever that day, but Harry's arrival was a special joy. He would understand exactly and he would think it was cool. I hadn't seen him for months.

We met in Berkeley in the late 1960s. We were both chemists. It was still ambiguous then as to how the synthesis of LSD was going to be tolerated in California. The law wasn't ambiguous. It was flat out illegal as of 1966 to synthesize LSD and a serious crime to even have it. But it seemed ridiculous that our legislature, which surely must have had important things to do with our limited tax dollars, would get itself involved in regulating LSD while scholars and popular magazines like *Time* and *Life* were still debating the pros and cons of this new phenomenon. A number of well-informed and respected psychologists were enthusiastic about the possible uses of LSD in psychotherapy, and various social and religious leaders saw LSD as a way out of World War III. But there was a lot of ignorance in Sacramento, California, about it, and wherever there is ignorance, you can always find arrogance. In Sacramento, arrogance was almost thick enough to rust the bumper off a truck.

We didn't believe they could possibly be serious. It was insane that the people who knew the least about something would be able to ban it. We figured the law would not be enforced. We were wrong on that one.

Harry and I both had a love of organic synthesis. The process of taking readily available things and turning them into precious substances is a little like cooking or magic. Harry had a love for larger-scale quantities than I did, so he was understandably secretive about his whereabouts after he left Berkeley. He left in the middle of the night after the campus cops had made an uninvited, and some say rude, visit to his lab.

The next time I saw him he was wearing a fake beard and talking about the fact that the *federales* were using voice printing technology to trace phone calls. I thought he was overestimating the cleverness of our national constabulary, whose skills, in my eyes, did not yet include Gerald Ford's famous ability to walk and chew gum at the same time. But I humored him. I made a few phone calls for him and picked up a package for him that had been hidden in his hasty retreat, and he rewarded me with a wonderful vehicle. It was my first old VW bus.

All that was long ago. When Harry came to my lab in San Francisco, without his beard, the timing was perfect. He saved me from the N-Judah streetcar, and BART, and Cynthia having to pick me up at the Fruitvale station in Oakland. We shot across the bridge in his new Toyota pickup, and by 1:00 A.M., we were sitting at my kitchen table with a couple of cold Becks.

Cynthia was asleep. Harry smoked a joint. I never smoked unless it was a weekend because I was older and marijuana made me too groggy the next day. Thirty doesn't seem old now, but it did then. Harry sipped his beer and then looked up at me differently, as though from across the aisle of some Buddhist bookstore, and said, "I want to show you something." I knew I was in for something good.

His eyes were wide, and he looked straight into mine. "Keep your eyes on mine and try not to blink."

I stared straight ahead. His face was calm, but his eyes were alive and intense. "If my face starts to change," he said, "don't react. Just keep looking into my eyes."

His face did change. It was still Harry, but varieties of Harry I had not seen. Different faces appeared out of the familiar flesh, which now wasn't so familiar. Some of them were humans I didn't know, some were not human at all. They were animal. They were all Harry in some way I couldn't explain. I was seeing things in him that were him but not a part of the life we had shared. It was a little scary, but Harry was somehow underneath it smiling that confident smile.

I trusted him more than almost anybody else I knew. He had experiences I wanted to share, and he didn't need anything from me that I was not willing to give. We had always had a nice balance that way. He had told me something a long time ago, when we were going into some business venture together, that made me always feel at ease around him. He said that for any human interaction to work both parties must believe they are getting the better deal.

It was hard not to blink, but I was totally enthralled and willing to concentrate in order to keep the images from going away. The kitchen started to take on a reddish hue, and the walls came in at the top and bottom like a barrel. It was a perception similar to one I had had after taking mescaline. I suspect it is a malfunction of the optical processing circuitry, which compensates for the fact that the images on your retina are really on a curved surface. Why I was experiencing a malfunction just then was not clear, but I was so distracted by an overwhelming sense of my own reality and my absolute permanence that I didn't pay much attention to the way things looked. I was aware that whatever I might wish or attempt—I could never not exist. I could not die. Rocks were fragile and could pass away, but not me. Nothing of me would ever go away.

Harry nodded. He understood what I was feeling.

"That's what I wanted to show you," he said. I was crying and so was he.

"I can read your mind, Harry," I said. "I'm only allowed into the front room—the reception area. But I'm in there."

"I'm in yours, too," he replied.

I got up from the table and broke the whole thing off for a moment. I knew for certain that all we had to do to re-enter that state of mind was to stare again into each other's eyes. I returned with a bunch of index cards and two pens.

"Write the next word you are going to say, Harry."

We were being scientists. We both wrote down a word and then showed each other our cards. It was the same word. Just a word, nothing cosmic, but it was the same, and we knew it would be. We did it again and again, and we knew every time it would be the same. We were watching something—always present but usually dormant—from a privileged position that we had created by putting ourselves together in some way. It was absolutely normal and yet it wasn't.

That night I recognized that whatever I had been experiencing and referring to as my life was only one aspect of something that was really me. That "me" was what people who were religious meant when they said "soul." It was nothing like the ghostly things I had imagined before whenever I thought about souls. Kary Banks Mullis was a ghostly thing compared to Kary. Kary was forever.

I remembered the slogan of the Bay Area Society for Life Extension. They were a bunch of crazies who were planning to immerse their freshly dead bodies in liquid nitrogen in order to be revived sometime in the future. The slogan had always seemed illogical to me. This morning it seemed absurd. "We want immortality and we want it now!" They hadn't the slightest clue about "immortality" or "now" or "we" or "want" or "it." I'm not sure about "and."

But I didn't know much about "and" myself until I met Nancy Cosgrove.

Harry and I did things like that on several other occasions. We were both impressed by what we had experienced, but neither of us was the type to fixate on things. Something I have noticed about other-worldly aspects to my life is that they don't tend to change a lot of what you had been doing already. They add something to what you might call the depth but not the direction.

8

INTERVENTION
ON THE ASTRAL PLANE

Two weeks before Christmas in 1978 I met Katherine O'Keefe in the flesh and began the final chapter of one of the most bizarre experiences that I've ever had. The story began in Kansas in 1974. One day, before doing my laundry, I decided to inhale some nitrous oxide, or laughing gas. I had a cylinder of it at home and liked to inhale it once in a while. I would breathe in a few breaths, and my mind would sail off briefly into something primeval and human-less. This time the effect would be very different, because the night before I had taken a powerful antihistamine with Cynthia Gibson. We were just getting to know each other—she would later become my third wife and the mother of our two boys—and she had been bitten by a mosquito at a backyard party I was having. She was allergic to mosquitoes and needed to take the antihistamine immediately, which would put her to sleep. She encouraged me to stay with the festivities, which were getting wild in a 1970s kind of way, but I tore myself away and joined her inside in an antihistamine reverie.

The next morning Cynthia went home to work on a paper for nursing school. I put the little plastic tube in my mouth and opened the valve on the tank. In my previous experience with nitrous I would have had plenty of time to react, turn off the tank, and then settle down for a couple minutes of bliss. The aftereffects of the powerful antihistamine of the night before changed everything. I was immediately out cold and dead to the world.

I woke up and the tube was on the floor in front of me. I had no idea how long I had been out. The gas was still running, it was cold enough to be condensing water out of the air, and the tube was frozen. The next thing I noticed was that my mouth felt funny. My tongue and lips were numb. I'd been anesthetized for a long time, and the tube was frozen solid. I shakily made my way to the bathroom, where there was a mirror. My upper and lower lips on the right side had bright white stripes from the frozen tube, and the tip of my tongue was white, like snow. Frostbite. The frozen tissue was melting, and it was starting to hurt. But much worse, there was going to be some hideous swelling.

I called Cynthia, who drove me to the hospital. My friend Marc was on duty in the pediatrics emergency room. I was starting to look really scary: my mouth looked like something out of a bad cartoon. I could barely say the word "frozen," and I wrote down the names of all the drugs I could think of that might reverse the swelling. Marc said he might have to do a tracheotomy. Was that all right with me? I told him I thought it would be swell—we laughed but we were scared. Marc started injecting epinephrine, norepinephrine, prostaglandins, antihistamines—everything we could think of that might cause blood vessels to contract and squeeze out the fluids converging on my throat. Norepinephrine worked. Within about an hour I was stabilized.

During the next month I made a miraculous recovery. The plastic surgeon I consulted had predicted at least a year of reconstructive surgery. Later she told me that she hadn't seen anything heal that well except in an infant. Well, I felt like one. It had been a dumb thing to do.

Cynthia was a wonderful nurse. She kept me alive with fresh juices and gazpacho through a straw. She read Dorothy Sayers to me in her third-floor bedroom. She scrubbed my wounds relentlessly with Dr. Bronner's peppermint soap.

I figured that I needed at least one cell that knew how to form the border between the pink and the white part of the missing edge of my lip. I hoped that cell was still alive somewhere. I was afraid it might be struggling to reproduce, and a powerful antibiotic like gentamycin might kill it. The plastic surgeon had advised taking it, but I didn't want to risk it. I imagined my mouth growing back. It was spring, and I was falling in love with my nurse. My lips and tongue grew back, and I needed no surgery. Within a month I could close my mouth completely and even whistle again.

There was a mystery to this story. Marc, the doctor who had treated me, was the first person to point it out. If I was unconscious long enough to have my tongue and lips frozen, how did the tube come out of my mouth? Animals anesthetized on nitrous oxide do not move. One of the advantages of nitrous in dentistry is that the patient doesn't wiggle or jerk around at all. The tank was half full, and there was still enough nitrous in there to keep me asleep for hours more. I should have been completely immobile, and I could have died. But when I awoke, the tube was some distance from my face. It was all very strange. What had happened?

Enter Katherine O'Keefe.

Cynthia and I were living in California. We'd gotten married and had two boys. My nurse turned out to be an understanding wife. She knew that I had a wanderlust that would be less trouble to deal with than to suppress, and she made a deal with me. Two weeks out of the year, right before Christmas and Easter, she would leave me to my own folly, after which I would join her and the boys at her parents' house back in Kansas. Whatever I did while she was gone would be over before she came back home. Keep in mind this was the 1970s.

It was Christmas of 1978 and Cynthia had been gone for six days. My week alone was almost over, and nothing of interest had happened. I had been hanging out at the Buttercup Bakery—not a likely place for me to come across a quickie romance because Cynthia and I used to work there together. But for some reason I was drawn to it. I was sitting at a table with two friends when Katherine O'Keefe came through the door. She walked directly to my table. I didn't know her, but our eyes connected, and within five minutes we were walking out the door.

She followed me home. We talked briefly about nothing much in the kitchen and then made love before I knew anything more than her name. She looked deep into my eyes and did something to me with her mind that was ecstatic. It seemed to me as if a little tentacle had reached into my midbrain and tickled my hypothalamus.

I asked her what the hell she had done to me.

She replied, "You've been playing with your mind, but you don't know anything yet. No one has ever properly taught you."

I was excited. "Will you show me how to do that? What you did?"

"You already know. You just need to practice."

Then she asked me if I had ever figured out who pulled the tube out of my mouth that fateful day in Kansas. My jaw dropped. No one except Cynthia and Marc knew about that tube. I hadn't talked about it. When you freeze your mouth by being totally stupid, you don't feel compelled to tell people about it. When I could finally speak again, I asked her how she knew about it.

"I was there, and I pulled it out of your mouth. I waited until I was sure you were okay and then I left."

I could hardly absorb this. How could she have been in my house in Kansas City?

It turned out that she could travel on the astral plane. Her mother had taught her how to do this when she was a child. It required that she imagine a machine surrounding her. The machine would respond to her intentions. She had been in transit when she had seen me dying. She knew I would later play a role in her life, so she stopped and pulled out the tube.

We talked through the night. I was absorbing what she said to me like a sponge. The next morning she woke me and reminded me that I had a plane to catch. I had forgotten all about my regular life and hadn't even mentioned it to her, but, of course, she knew.

After Christmas I made arrangements to drop by Kathy's house on the way home from work on Thursday afternoons. She had agreed to teach me what she knew. She told me that I had abilities that I hadn't tapped into and that I had to learn to quiet myself inside. I had to learn not to think so much.

I was not unaware of meditation, but I had always found it difficult to do in the way it's usually done. I spend a lot of time in my meditative states, but I didn't know how to describe it to Kathy. So I struggled. Kathy didn't realize that what she needed was not to teach me something new but to teach me to use what I already had.

I started being able to do little exercises with my mind. Nothing fantastic, but I was a willing and eager student. I wanted to know what it would be like if I was in the room with her at the same time as she was traveling by her imaginary machine to New York to be with her mother. She didn't always answer my questions. Sometimes she looked at me as if I shouldn't have asked them, but that didn't stop me. I've never been of the opinion that any question is off-limits.

On the first Thursday in March I came by at the usual time; she met me at the door.

"You have to stop coming by. Something bad has happened and I need to concentrate. You're too distracting for me. I need to find someone—like me—to work with."

It was final. There was suddenly no space in her life for me. I went away and I didn't hear from her again until the fall.

I was now working at Cetus. Katherine wanted to meet me for lunch. I saw her coming down the sidewalk and walked toward her. Something felt very ominous.

"Do you remember when I told you that I would need you someday?"

I had almost forgotten.

"Well, I need you now. I'm about to die, and I'm not ready. My children need me."

Kathy had discovered back in March that she had a malignant melanoma. She had "found somebody who was like her," and the two together had tried to cure her, but the cancer was spreading. She was dying. She wanted to know if I could get her interferon at Cetus.

We had about ten micrograms of it there as a reference. Cetus was eventually going to produce it in amounts that would be therapeutic for some types of cancer, but it was years away from clinical trials. I had to tell her that I couldn't help her. I couldn't even delay her death.

"I guess I knew that before I called you, Kary. It was just a dream."

I asked her whether there was anything I could do for her. She said no. Two weeks later some guy called me to say that Kathy was dead and she had wanted me to know.

9

AVOGADRO'S NUMBER

IN THE EARLY years of PCR, no one could figure out why certain methods of doing it turned out to be better than others. As I first envisioned PCR, each cycle would cause the amount of target DNA to double. The first cycle would provide twice as much, the second cycle four times as much, the third cycle eight times. But often by the tenth cycle, it would not completely double. It would increase by a factor of 1.8, then 1.5, and 1.3. Something was running out or something was being made that interfered with the process. It meant that calculations based on a consistent doubling could be way off. Like compound interest with a variable rate.

In PCR there are about twenty different things you have to measure out, each of them dependent on all the others. In figuring out how to perform the reaction so that each cycle would result in a complete doubling of the target molecule, a major problem I encountered was that I had to deal with several completely different systems for measuring the amounts of things. Some of the molecular ingredients were measured in grams, some of them in esoteric units of activity, and some of them by how much ultraviolet light they would absorb. The reasons for all this were historical and constituted a real pain in the science of chemistry that no one had taken the trouble to fix.

I decided to fix it.

The most rational of the systems of measurement was due to a concept put forward by a nineteenth-century count named Amedeo Avogadro. He taught higher physics at U.T. While at Turin, he suggested that equal volumes of gases contained the same number of gas particles. Sometime later that century, Stanislao Cannizzaro used Avogadro's concept to devise a system of chemical measurement, which is still in use today.

Chemistry students hate it. Chemistry adepts love it because it separates them from the students. Before doing a chemical reaction, which is like a recipe, you have to determine the amount of each ingredient you're going to use. If every molecule weighed the same, that wouldn't be difficult. But because the molecules in each chemical have a different weight, this gets to be very complicated.

Every chemical has written on the front of the bottle its molecular weight, which is defined as the number of grams of the chemical that contain Avogadro's number, 6.023×10^{23} of molecules. That number is called a mole. For example, one mole of carbon weighs 12.011 grams, whereas one mole of glucose weighs 180.16 grams, but the number of molecules in a mole of carbon is equal to the number of molecules in a mole of glucose.

If this seems complicated, well, it is because 6.023×10^{23} is an astronomical number. Chemists like the fact that nobody can really understand what they're talking about. Besides the difficulty in computing, the problem that they have failed to see is that in using this system of nomenclature they are making things so complex that they themselves miss out on obvious, glaring mistakes. That was a problem I had in trying to understand why my PCR reaction was so inconsistent.

Avogadro's number has no inherent chemical significance. In many calculations, I noticed, biochemists were first multiplying by Avogadro's number, then somewhere later in the same calculations were dividing by this number. It's easy to make huge mistakes when calculating numbers with exponents like 10^{23}. Norman Arnheim, whom I worked with at Cetus, did a long calculation to show how many moles of DNA would be present in a single sperm cell. He screwed up the calculation and ended up with 0.9 molecules instead of 1 and published a paper including the hilariously erroneous conclusion.

I began to wonder how we managed to create such a confusing situation. In Europe the gram is a commercial unit. People buy bananas by the gram. But in America the gram is primarily a chemical measurement. The gram unit of weight is based on the fact that a cubic centimeter of water weighs 1 gram in its most dense state, which is at 4 degrees centigrade. The centimeter is 1/100 of a meter. A meter, according to the French, is 1/10,000,000 of the distance from the North Pole to the equator on a line that went right past Notre Dame, where that little guy sells roasted chestnuts in the winter. For chemistry the measurement is inappropriate because it's a geophysical concept.

When chemists actually did mix 200 grams of one substance with 500 grams of something else in a big flask, Avogadro's number was marginally useful. But chemists rarely deal with anything that large anymore. Rather than 200 grams, people are mixing substances on a molecular level—200 molecules of this and 200 molecules of that. Instead of dealing in liters, we are using microliters. Chemists can now work with two molecules; they can detect them and do reactions with them. The old system just isn't applicable anymore. What chemists need to be able to determine is how many molecules there are someplace and how close together they are. There's a very simple way to determine that number: Count them. When you have ten units of an ingredient, express it as ten. It isn't necessary to say ten divided by some unwieldy number.

I created an entirely new system for comparing substances, the details of which would be of considerable interest to only a small group of people. But it was by using this system that I was finally able to figure out why people doing PCR were getting inconsistent results. When I compared the number of molecules in each of my ingredients, it became obvious that in many cases there simply weren't enough enzyme molecules to react with the DNA molecules that had to be processed. Nobody had realized that the limiting factor in doing PCR is simply the number of molecules of the enzyme, because they had no way of knowing how many molecules of enzyme they were using.

When I started experimenting with PCR, I never knew how many molecules of enzyme I had in my solution. Enzyme amounts are expressed in units of activity, how many molecules of something it will turn into something else, under a standard set of conditions per unit of time. Only after I converted each of the ingredients needed to do PCR to a simple system of counting molecules did the problem, and the solution, become obvious. Keep the number of things the enzyme is going to interact with smaller than the number of molecules of the enzyme. Simplicity is embarrassing when you have to work for months to achieve it.

10

WHO'S MINDING
THE STORE?

WHEN WE WERE children, we thought our parents were taking care of things. Sometimes they were. As adults, we like to think that there are some very wise people, usually older than we are, taking care of the planet and us. As a result of this wishful thinking, a lot of people make a living under the pretense of doing just that.

It would be naïve to think that individuals working in government agencies charged with taking care of us, or even in nonprofit foundations with lofty names, are altruistic toward us. They aren't sharing our genes. They aren't our parents. They are attending to their own biological imperatives and their own personal needs. Only when "ours" and "theirs" overlap do we get attention.

Once in a while—wartime, for example—we all pull together. The general and the troops all have an equal and obvious stake in avoiding annihilation by the enemy. In peacetime it doesn't happen often that strong pressures for individual biological success—life, liberty, and the pursuit of happiness—overlap with pressures for survival of the group.

No one is looking out for our best interests. Not the church, not the president, not even Mother Teresa—Christianity,

Green Peace, and all the other Green Things notwithstanding. We're on our own as always.

This is not new with the twentieth century. The constitutional government of the United States of America was set up with the notion of having checks and balances in government. The framers of that document were practical and aware that we cannot count on having philosopher kings or presidents who always act in the best interests of the country. We need two or more governments working in parallel, competing for control within a civil system to prevent a government from getting out of control and having to be displaced in an armed insurrection. It's the best we can do under the circumstances, and it has worked pretty well. But there are new problems.

What has happened in this century is that the world has become increasingly complex. Many functions of government have spread into highly technical areas that are impossible for concerned outsiders to monitor continually.

The National Institutes of Health is one such monster. The Environmental Protection Agency is another. The National Oceanic and Atmospheric Administration hires people who advance their careers by telling us about the hypothetical effects of sulfate aerosols, as though there were a real, scientifically sound connection between sulfate measurements and the weather in the next millennium. The Patent Office is another bureaucratic mess. The Federal Reserve Board is a tawdry sepsis. No one who works there has to worry too much about interest rates.

How can we bring the spirit of checks and balances into the massive arms of an enormous bunch of faceless bastards working, or sometimes just enriching themselves, doing God knows how many technical tasks?

When Congress passes a law that is not in keeping with the contemporary interpretation of the Constitution, the Supreme Court usually understands what is up and overrules it.

When the National Institutes of Health makes an announcement through one of its many spokespeople, who checks out the credibility of that statement?

Checks and balances are hard to come by in a scientific establishment that is supported from outside by a populace unskilled in the scientific arts. I know it's going to be a hard and inefficient answer. Compared to a benevolent monarchy, having three branches of government was also inefficient. And I know that as long as it achieves a better life for us here in the colonies, we will put up with it. We are optimistic people really, and we are not in a hurry to go anywhere else. I don't know exactly what the answer is, but I know that the answer is not to believe, "Trust us. We're here to help." It never has been.

In my naïveté, the world was a safe place until 1968. I thought it was watched over by an elite group of people with great wisdom who had proven themselves and were entrusted with protecting us and the planet. I hoped that I, a conscientious twenty-two-year-old who loved to learn and teach, would someday be a member of that group.

In the early weeks of 1968 I submitted an article I had written to the foremost scientific journal in the world, *Nature*, published in London. I called it "The Cosmological Significance of Time Reversal" and congratulated myself on its cleverness. It was a description—from my own experience and imagination—of the entire universe from the beginning to the end.

It was one of those intuitive things that needed to be expressed as a tentative hypothesis, on account of my limitation experientially to the *right now* and my somewhat limited experience as a cosmologist. I was a second-year graduate student in biochemistry at Berkeley. I had read a lot about astrophysics and had taken some psychoactive drugs, which enhanced my perceived understanding of the cosmos. Not very good reasons to think that an international journal of science would want to publish my views for the edification of their very knowledgeable readership.

It was accepted. I received a flurry of letters from all over the world requesting reprints. At first I was elated by the response. Nature Times News Service circulated an article beginning, "It sounds like the wildest science fiction. But an American scientist seriously suggests that half the matter in the universe is going backwards in time." Some lady in Melbourne sent it to me with a letter asking for my autograph. Later in the article they referred to me as "Dr. Kary Mullis of California University." I began to be a little concerned. Something was definitely amiss in the world of science.

I was not a doctor. I was still a student, only hoping to become a doctor. Who had promoted me to doctor? Why would the news services pick up the story and print it all over the world in the papers? I was not really an experienced astrophysicist. What did I know about the universe?

I grew up. I lost that long-abiding feeling that there were older, wiser people minding the store. If there had been, they would not have allowed my first sophomoric paper on the structure of the universe to be published in the foremost scientific journal in the world.

Years later I invented the polymerase chain reaction. I was a professional scientist, and I knew what I had discovered. It was not the speculations of a kid about the universe and time reversal. It was a chemical procedure that would make the structures of the molecules of our genes as easy to see as billboards in the desert and as easy to manipulate as Tinkertoys.

PCR would not require expensive equipment, and it would find tiny fragments of DNA and multiply them billions of times. And it would do it quickly.

The procedure would be valuable in diagnosing genetic diseases by looking into a person's genes. It would find infectious diseases by detecting the genes of pathogens that were difficult or impossible to culture. PCR would solve murders from DNA samples in trace materials—semen, blood, hair. The field of molecular paleobiology would blossom because of PCR. Its practitioners would inquire into the specifics of evolution from the DNA in ancient specimens. The branchings and migrations of early man would be revealed from fossil DNA and its descendant DNA in modern humans. And when DNA was finally found on other planets, it would be PCR that would tell us whether we had been there before or whether life on other planets was unrelated to us and had its own separate roots.

I knew that PCR would spread across the world like wildfire. This time there was no doubt in my mind: *Nature* would publish it.

They rejected it. So did *Science*, the second most prestigious journal in the world. *Science* offered that perhaps my paper could be published in some secondary journal, as they felt it would not be suitable to the needs of their readers. "Fuck them," I said.

It was some time before my disgust with the journals mellowed. I accepted an offer by Ray Wu to publish it in *Methods in Enzymology*, a volume he was preparing. He understood the power of PCR.

This experience taught me a thing or two, and I grew up some more.

No wise men sit up there, watching the world from the vantage point of their last twenty years of life, making sure that the wisdom they have accumulated is being used.

We have to make it on the basis of our own wit. We have to be aware—when someone comes on the seven o'clock news with word that the global temperature is going up or that the oceans are turning into cesspools or that half the matter is going backward—that the media are at the mercy of the scientists who have the ability to summon them and that the scientists who have such ability are not often minding the store. More likely they are minding their own livelihoods.

11

WHAT HAPPENED TO
SCIENTIFIC METHOD?

J AMES BUCHANAN ADVANCED an ugly idea that got him a Nobel Prize in 1986. Buchanan cannot be held responsible for the ugliness; we can't blame the messenger. It came to be known as public choice theory. You will recognize it and wonder why people get Nobel Prizes for pointing out such simple things. The answer is that most people can't see the simple things and the simple things are always the most important.

Buchanan divided the world into four groups—voters, politicians, bureaucrats, and interest groups. Everyone in each of these groups wants something from the System, and everyone but the voters are organized professionals. The voters have to go to work every day. They cannot concentrate from nine to five on how to get something from the System. Most of us fall into the voter category.

The chairman of the Federal Atmospheric Commission is having dinner tonight with his secretary and some expert they flew in from Cal Tech. He just announced tonight on CNN that his lab needed another $50 million dollars to study ozone depletion. Over swordfish, with the new science correspondent from CNN, they are congratulating themselves on what a smooth job they did today on the news. Crème caramel and

B&Bs and cappuccinos later they promise to meet again in Oslo at the Envirocon World 2000 meeting next fall.

The ozone story is on NPR as you drive home from work. You feel terrible about all those years that your shaving cream came from an aerosol can, and your wife was using aerosol hairspray. You seem to be feeling guilty a lot these days— almost every time you turn on the news and hear about the environment.

You stop by Walgreen's to get some shaving gel without CFCs. The large pepperoni pizza in the back seat stinks up the car and it's getting cold, but you've done your best for the planet. The family isn't impressed when you get home. Maybe you think it would be nice to have some B&B of your own tonight, but it's April. Taxes are due, and your wife, watching *Seinfeld*, doesn't want to talk about it and wouldn't think it was safe for you to go down to the liquor store this time of night anyhow. Your daughter reminds you that you haven't sent the check to Greenpeace, and by the way, she's definitely going up the coast this weekend to the protest over the Marin Headlands Interior Department deal, and your wife says she can't, and "Would you please talk to her, Dad. She acts like I'm going up there for fun."

Is this you? Or were you the one with the mushrooms and red peppers?

Who are these people who make comfortable salaries arranging scientific symposia and stories for the media? They aren't politicians. Politicians don't know anything about scientific things. They just want to look like they do. Somebody has to advise them. Who are those advisors? It's an important question because those people—who are always having to come up with the imminent disasters that can be prevented by governmental projects, sponsored by informed and well-meaning politicians—are manipulating you. They are parasites with degrees in economics or sociology who couldn't get a good job in the legitimate advertising industry. They are responsible for a lot of the things that you accept year after year as your problems. The problems they imagine for you are as imaginary as the commercials during *Seinfeld* about some Australian outback macho guy, with a Hollywood model by his side, driving a four-wheel-drive vehicle, with pathetic half-wits in pursuit due to a misunderstanding about the relative merits of the vehicles.

Who pays these experts? Is it the Intergovernmental Panel on Climate Change that the United Nations is supporting with our money? Or is it the Environmental Protection Agency, which you were bitching about today because your company was having to close down one of its plants due to some fish that might go extinct, and you might get transferred in the shuffle? Is it the Tropical Oceans and Global Atmosphere Group? Is it the Arctic Climate System Study? Is it the Marlowe Walker Eternity Endowment? Is it the World Ocean Circulation Experiment? Is it the World Bank's Global Environment Facility? Is it Greenpeace? The Sierra Club? You are too tired from your day at work to try to figure it out. That's what James Buchanan predicted. But the sun never sets on the British Empire or bureaucrats—environmentalists, as many of them are called today. Sleep soundly. Your planet is in well-fed hands.

Now, I like to hear a good story. I like to tell one. But when my car isn't working, I want to know why—in terms that I can understand. I don't want to fix it myself, but I'm more comfortable if I understand what the problem is. I don't like it when some mechanic, looking at my clean fingernails, thinks he can entertain me for a minute with conversation about the modern features of my car and then sock it to me. I feel the same way about our planet and my food.

If there is something in my food that somebody says is a poison, I want to have the chemistry explained and decide for myself whether or not I want to eat it. Science is a method whereby a notion proffered by anyone must be supported by experimental data. This means that if somebody else is interested in checking up on the notion presented, that person must be allowed access to instructions as to how the original experiments were done. Then he can check things out for himself. It is not allowable in science to make a statement of fact based solely on your own opinion.

Claims made by scientists, in contrast to claims made by movie critics or theologians, can be separated from the scientists who make them. It isn't important to know who Isaac Newton was. He discovered that force is equal to mass times acceleration. He was an antisocial, crazy bastard who wanted to burn down his parents' house. But force is still equal to mass times acceleration. It can be demonstrated by anybody with a pool table and familiar with Newton's concepts.

Science appeared in the seventeenth century.

It was not the first time that humans had ever done science. The pyramids at Giza in Egypt strongly suggest an earlier scientific age. There are manuscripts, for instance, the texts of Euclid, which were translated into Arabic by clever Arab scholars while ignorant Arab soldiers were destroying the greatest library in the ancient world at Alexandria. Some of the translations luckily ended up in Toledo, Spain, where they sat on top of a steep, defensible hill awaiting the eventual liberation of Spain and the reclamation of some of the glory of early Mediterranean culture by the French, who were some of the descendants of the poor people who had fled when the whole early Western thing had collapsed in Italy hundreds of years earlier. Or something like that. The details are mostly lost. But eventually science did reappear.

Robert Boyle, who was a Christian and a friend of the English monarch Charles II, made a vacuum pump in the seventeenth century and showed that he could extinguish a candle by pumping the air out of the jar wherein the candle was burning. According to Boyle, whatever was left in the jar after the candle went out constituted a vacuum. In the common vernacular, it meant that absolutely nothing was there. Whether God was in there or not was not something Boyle addressed. He didn't know how to measure the existence of God. The religious issue was not as interesting as the issue of what he could measure. The Catholics seriously disagreed. They had documents which clearly stated that God was everywhere. Even some garbage from mistranslations of Aristotle that said "Nature abhorred a vacuum" was taken to mean that Nature just fucking wouldn't allow one at all and that Boyle was an idiot. But the candle went out. Boyle didn't care whether God was there or not because he couldn't measure God. That's when science started to take off.

Computer modelers of the ozone layer and the next thousand years of climate could take a lesson from Sir Robert Boyle and his Royal Society. If you can't actually measure something, or make an accurate prediction from a theory, and present it to a group of your fellows, be good enough not to disturb us about it.

Boyle realized that we're living in a fluid. Fish probably don't realize they are living in a thick, viscous fluid. They're born in water, it's a constant in their habitat, so they aren't aware of it. They might call it nothing, and if they started ascribing philosophic or religious properties to their misconception they would run into problems. We thought we lived in nothing, but Boyle showed that we live in air and what is left when we pump it out is a vacuum and that is something different from air—even though it looks the same.

People who accepted the existence of a vacuum gave their allegiance to the king; people who believed the creation of a vacuum was impossible supported the pope. In 1662, Charles II chartered the Royal Society of London for the Improvement of Natural Knowledge. Boyle was one of the founding members. Those interested in scientific discovery were invited to the Royal Society to demonstrate how things worked. It was through use of this scientific method that science was separated from religion and philosophy, and that included morality. Science freed of morality began to shine.

The laws of science are demonstrable. They are not beliefs. When experiments in our century showed that Newton's gravitational laws were not quite accurate, we changed the laws—despite Newton's good name and holy grave in Cambridge. Relativity fit the facts better. This is the way science has been done now for almost four centuries, and because of science—not religion or politics—even people like you and me can have possessions that only a hundred years ago kings would have gone to war to own. Scientific method should not be taken lightly.

The walls of the ivory tower of science collapsed when bureaucrats realized that there were jobs to be had and money to be made in the administration and promotion of science. Governments began making big investments just prior to World War II. Scientists and engineers invented new firearms, sharper things, better engines, harder things, airplanes that could fly faster, radar to detect them, antiaircraft guns to shoot them down, antibiotics for the pilots who got shot down, amphetamines to keep everybody awake long hours, daylight savings time to lengthen the hours, and finally one big bomb that in a shocking finale brought World War II to a breathtaking and hideous end.

Scientists had revealed that they weren't just a bunch of screwballs who had nothing to do with the world. They were not, and never had been, useless little guys sitting in ivory towers playing with slide rules. Just a few of them, with motivation and some tools, could make a bomb that would have put the fear of the Christian God into Attila the Hun.

Science was going to determine the balance of power in the postwar world. Governments went into the science business big time.

Scientists became administrators of programs that had a mission. Probably the most important scientific development of the twentieth century is that economics replaced curiosity as the driving force behind research. Academic, government, and industrial laboratories need money for salaries for staff: the primary investigator and his technicians, postdocs, graduate students, and secretaries. They need lab space, equipment, travel expenses, overhead payments to the institution, including the salaries and expenses of administrators, financial officers, more secretaries, maintenance of grounds around the institution, security officers, publication costs for scientific reports in scientific journals, librarians, janitors, and so on. It's expensive, and there is a lot of pressure on a professional scientist trying to maintain or expand a laboratory domain. Most of the money comes from institutions like the National Science Foundation, the National Institutes of Health, the Defense Department, and the Department of Energy. There is serious competition for these funds. And the question we should ask is, "What the hell are you doing with our money that is so important to us?"

Imagine two hypothetical labs competing for public funds.

One of those labs announces in a series of scientific papers that they have found some unexpected and very interesting phenomena in the upper atmosphere that contradict the currently accepted theories on the radiogenic formation of carbon-14. This could have a dramatic impact on the radioisotopic dating of fossils. The time frame for human evolution might be a tenth of what has previously been concluded. We may have evolved from the fossils in the Oldavai Gorge in only a couple of hundred thousand years. All of biology may be much younger than we think. More research would be required to confirm this. Biologists all over the world are curious and very excited. The lab is requesting a million dollars from the National Science Foundation to conduct a more detailed study.

A second lab working on upper atmospheric physics calls a press conference to report preliminary data on what appears to be a giant hole in the ozone layer and warns the reporters that if something isn't done about it—including millions of dollars in grants to study it further—the world as we know it will be coming to a tragic end. Skin cancer is epidemic, and there are reports of sheep going blind from looking up to the sky. People are starting to worry about having sunglasses that shield their eyes from ultraviolet light. Children begin to learn about it in school, and they are taught to notice the intensity of the UV light when they get off the bus.

Which one of these two laboratories will get funding? Follow the money trail from your pocket to the laboratories and notice that it passes through politicians who need you and by the interest groups who with the media train you. James Buchanan noted thirty years ago—and he is still correct—that, as a rule, there is no vested interest in seeing a fair evaluation of a public scientific issue.

Very little experimental verification has been done to support important societal issues in the closing years of this century. Nor does it have to be done before public policy decisions are made. It only needs to be convincing to the misinformed voter. Some of the big truths voters have accepted have little or no scientific basis. And these include the belief that AIDS is caused by human immunodeficiency virus, the belief that fossil fuel emissions are causing global warming, and the belief that the release of chlorofluorocarbons into the atmosphere has created a hole in the ozone layer. The illusions go even deeper into our everyday lives when they follow us to the grocery store.

People believe these things, and a slew of others, not because they have seen proof but because they are ingenuous: they have faith. These issues don't have to be on faith. They are not transcendental. Some of them are hard to investigate, because you can't do experiments easily with people's daily lives, but they can be investigated, then confirmed or dismissed. If not, scientists should not be talking about them. Newton would not have allowed someone to carry on about saturated fats and heart attacks inside the Royal Society because like so much of the nutritional garbage that we are assaulted with daily, it is all conjectural, awaiting further study that will probably not be done.

Scientists who speak out strongly about future ecological disaster and promote the notion that humans are responsible for any changes going on are highly suspect. Turn off the TV. Read your elementary science textbooks. You need to know what they are up to. It's every man for himself as usual, and you are on your own. Thank your lucky stars that they didn't bother to change their clothes or their habits. They still wear priestly white robes and they don't do heavy labor. It makes them easier to spot.

"Ecological" is a word like "universe." It doesn't mean anything really. It is relevant because relevance is totally subjective, totally subject to public whim, and everybody now thinks ecology has ultimate relevance. Taken out of the context of conservation of the present situation, what does "ecological" really mean? It collapses back to the politically less-motivating Smokey the Bear. Emotionally it loses its hold on us if we are willing to look honestly at the history of the planet we so love and notice that the thing that is absolutely constant here is serious change—uncomfortable, sudden, cataclysmic change. What is the trouble with something being out of balance if the natural state of that thing is change? Who came up with this hallowed idea of ecological balance?

I couldn't help but notice the amazing coincidence that the

American patent on the production of freon, the principle chlorofluorocarbon used in refrigerators and air conditioners, expired at just about the same time freon was banned. Those countries that had begun producing freon without paying for the privilege were asked to stop. And a new chemical compound, a commercial product that would be protected by patent, would soon be substituted and make a lot of money for the company that produced it.

Indirect evidence pointing to a decrease in the ozone layer is absurd. There has been an increase in reported skin cancers. Reports of skin cancer may be increasing, but that isn't a good indication of UV levels. Increased skin cancer might have been caused by people moving to sunnier climates. People from America's North and Northeast have moved to the South and Southwest in the last forty years. During this same period, suntans became a fashion statement. Why not blame the increase in skin cancers on golf? It also might be that doctors and their patients have learned recently to look for those little fast-growing dark spots on the skin and have simply gotten better at diagnosing and reporting skin cancer. To measure the amount of UV reaching the Earth unambiguously, you would not measure cancer, you would measure the UV light reaching the earth. Just put a $6,000 UV measuring instrument on the ground at one of those stations in Antarctica and check it for a few years. Couldn't somebody do this and report it? If they have, I haven't heard about it.

Beyond the lack of scientific evidence, it makes no sense anyhow that we could destroy ozone in the upper atmosphere. If a hole in the ozone layer appeared somehow, here's what would happen: The UV rays from the sun would come through that hole and strike the Earth's atmosphere, where they would be absorbed by the miles-thick layer of oxygen surrounding the Earth. Then it would make more ozone. When the UV rays from the sun combine with oxygen, they form ozone. The ozone thus formed absorbs UV light, which continues to come from the sun, and prevents it from penetrating any farther into the oxygen below that has not been converted to ozone. That is why we have oxygen to breathe down here and ozone in the upper atmosphere. If all the nations in the world agreed to spend all of their money to eliminate the ozone layer—they couldn't do it. It can't go away unless all the oxygen in the atmosphere were to go away, and then, guess what—we couldn't breathe, until the green plants made some more. The ozone in the upper atmosphere regulates itself. If you measure a drop in some variable like ozone, it doesn't mean it is going or gone. Put a stick on the beach marking the edge of the last wave while the tide is coming in, then come back in an hour with another stick. You'll notice that the tide has come in ten feet in an hour, but if you predict that in a year the tide would have come in 87,000 feet, you'd be dead wrong.

The concept that human beings are capable of causing the planet to overheat or lose its ozone seems about as ridiculous as blaming the Magdalenian paintings for the last ice age. There is a notion that our emissions are causing the temperature of the planet to go up, even though the temperature is not going up. Even if the temperature were going up, we would be foolish to think we caused it. We could just as reasonably blame it on cows. In the nineteenth century the temperature went down. In this century it's gone up only about half a degree. The trend over the last two centuries is down. Down is not warmer. So if you like to worry, worry that we might be moving into a new ice age. We could be.

Would that be something we would want to stop? We didn't cause the last ice ages and we didn't cause them to go away. We benefited from them. We don't cause thunderstorms and lightning either. We don't cause the El Niño years anymore than we cause the other years. We don't cause floods. We live on a planet that has many mysteries, including the patterns of its changing climate. We are the children of those changes, and we derive from those mysteries.

We accept the proclamations of scientists in their lab coats with the same faith once reserved for priests. We have asked them to commit the same atrocities that the priests did when they were in charge. We have forced this situation by requiring that they bring us relevant innovations. We have turned them into something almost as bad as lawyers. Something to toy with us and our strange needs. Scientists could be something to entertain us and invent nice things for us. They don't have to be justifying their existence by scaring us out of our wits. Can't they be comforting? It's up to us, not them, because they depend on us for support. We have to arrange them in such a way that they and we benefit from the arrangement.

Hundreds of years after Boyle's experiments, we still haven't learned to separate matters of fact from our beliefs. We have accepted as true the belief that we are responsible for global warming and a growing hole in the ozone layer—without scientific evidence. We have faith in disaster. Scientists have a considerable financial stake in our continuing to believe that these problems threaten our lives and must be solved. They get paid for it. What do we get out of it? Is it a feeling of comfort, of knowing that our lives are being protected?

Perhaps the best solution for our anxiety is to do exactly what our ancestors did. Build some churches in the Gothic style. Fill them with nice art. I like pictures in bright colors of stern-looking people with halos, but whatever works is okay. Bring artisans from Sweden to build pipe organs and sponsor composers from Germany, Poland, England, and New Orleans to write some hymns, castrate some young boys for the descant parts, and come every Sunday to sing together and pray for our souls. Keep the Freon. We'll need the churches to be air-conditioned in the summer.

12

THE ATTACK OF THE
LOXOSCELES RECLUSAE

I LEARNED WHAT would happen if I put my hand on top of a red ant hill. I was curious to know how it felt. I knew that if I kept my hand very still, the ants wouldn't react. Ants don't bite just for the hell of it. I neglected to consider how I would go about getting them off my hand. I should have had a bucket of water nearby to put my hand into when I was through with my experiment. Ants will float when immersed. Instead, I started scraping them off. Every ant on my hand decided it was time to bite. When I got home that afternoon, my hand was painfully swollen. My mother's advice: "Don't play with ants, Kary."

After that, I was more careful, but I never stopped playing with ants.

My brother Robert and I would put insects together in a mason jar to see what would happen. We found out that if you put a black widow spider and a hornet in an enclosed jar, they go at it right away, and the hornet wins. If there is no lid on the jar, the hornet flies away.

A praying mantis is a fascinating pet and easy to catch. It is fun to watch her creep up on a fly and eat it. The praying mantis is delicate and masterful and quick. The fly scarcely knows what's got him. The mantis starts with the head so as to enjoy the meal without distraction and ends with the wings. Probably the dry, scaly wings are the worst part. But maybe, considering the flourish with which she finishes them—stuffing them in her rotating mandibles like a Frenchman savoring the last of a good Cognac—I wonder whether the wings aren't possibly the best part.

She then cleans her mouth parts with her legs, and her leg parts with her mouth. Very civil.

I've heard, but never observed, that a female mantis will do something similar to a male mantis whom she has lured to her side. She starts with the head. The decapitated male body, in spite of losing his head, still does what all male bodies will do if given the chance. Secure in the knowledge that she will lay fertile eggs, she finishes off the libidinal feast.

Once I gave a caged mantis a huge South Carolina moth. I opened the cage door when they started banging around against the cardboard. The moth flew, towing the mantis under its belly like a 747 taking the shuttle back to Cape Canaveral. The mantis swelled up like a balloon and looked like it might burst. It was eating the moth. I captured the moth again in one of its low swoops and pulled the mantis off. I put it back in the box, and the next day it was dead. Probably of frustration.

I T WAS 1996, and I was having lunch with David Fisher. "What's that black stuff on your elbow, Kary?"

I lowered my elbows. There were two black spots on my right elbow. Tar black. They looked like scabs, only they were too black and too round. I knew I hadn't scraped myself.

"I don't know." I picked at my new body parts tentatively. They were still attached, and we were at a nice Italian restaurant in Berkeley where people don't lift off scabs. We dropped the subject.

After lunch I drove north on 101 to Mendocino. I kept checking my right elbow. The skin felt tight. The blackness was getting darker and the roundness wasn't turning elliptical or smeared. Something odd was going on.

That evening, the first scab broke loose and revealed a shallow pool about a centimeter wide filled to the brim with my white corpuscles. It didn't look like a healing wound. The pool seemed to be seething with life, and my elbow was noticeably warm.

I checked out the *Merck Manual,* a reference book that no cabin should be without. Years ago I had been frightened by an exploding capillary in my eye. It had appeared as a pinhead of blood under the layer of eye skin called the conjunctival membrane and it had spread under the membrane across the white of my eye in the gruesome redness that only blood can express. The *Merck Manual* had calmed me down. The book said it happened once in a while and was not an indicator that it would happen again. The worst part was that it looked scary.

This time the *Merck* was not so comforting. I seemed to have been in the company of *Loxosceles reclusa,* the brown recluse spider. The manual impersonally advised that I was in for some serious shit.

I had seen brown recluse bites on people's faces in medical books where the bites won and the faces lost. Thank God I only had two on my elbow. The black scab appears about twelve hours after the wound and falls off in another six. I must have been attacked in La Jolla. I was 600 miles from there now. I went to sleep feeling safe. I didn't know that the northern California branch of the *Loxosceles* family, in touch with their southern California brethren via my suitcase and Southwest Airlines, was waiting for me in my greenhouse/bedroom. I guess I smelled like spider pizza.

The Mendocino recluses had their way with me. By morning I had eight new spider bites. They call them bites, but they are really excavations. Spiders don't have teeth and they don't bite. They had worked the old wounds from La Jolla—drinking my fluids and shooting in a little more venom to improve the flow rate. Then they dutifully punched a few more holes. The new wounds developed quickly. Well-nourished spiders make lots of venom. The venom, gently injected through the fangs so as not to damage them, kept me from making an immune response to the skin bacteria that had been scraped into the hole. The bacteria have names like Staphylococcus and my body knows how to defend me against them, but not in the presence of brown recluse venom. Without my immune system the staph can live happily on my epidermal cells. And they can go deeper. It becomes a dermis party. Staph will turn my skin into mush: spiders need liquid food.

One of the new holes was on the left side of my nose almost in my eye. I was very worried about that one. The *Merck Manual* suggested that I use hydrogen peroxide on the wounds, but it offered no real good news. It suggested that surgery might be a good idea and that the knife should go deep. A shallow excision of the wound would simply result in another, deeper, still necrotizing wound. Necrotizing means dying flesh, an expanding hole making more pus.

Jesus, I thought, pondering the effects of deep scalpels. I took some Vicodin for the pain.

On the Internet you can find a description of the brown recluse from the Nebraska Institute of Agriculture, which says without qualification that "spiders attempt to bite humans only as a last resort when threatened, injured or trapped in clothing. They prefer to retreat rather than attack and will generally avoid contact with humans."

Last resort? Threatened? I was asleep in my own bed.

The University of Kentucky on the Internet naively underestimates the evil that lurks in the spider heart. "The brown recluse roams at night seeking its prey. It is shy and will try to run from a threatening situation but will bite if cornered."

The fuck it will. It doesn't bite; it has no teeth. It scrapes with the tools on its front legs, and whether it's shy or not is irrelevant. It takes you in your sleep. You're not embarrassing it with personal questions.

It's a mother spider that first gets you and she wants a hole in you that oozes and expands and doesn't ever heal. The females have the most powerful venom, according to the experts from the University of Kentucky, College of Agriculture, Department of Entomology. She wants that hole because her babies need a place to feed. They can dip their ugly little heads into the pool of nutrients that you are exuding and suck your vital fluids through their sucking tubes, and they can live. Nobody's threatening the spider. After making the hole, she moves away and lays her eggs. It's elegant biotechnology from the point of view of the spider.

For the human, it's a flagrant disrespect of personal sovereignty. It's an unnecessary surgical procedure being undertaken without permission of the patient or the next of kin. It's a hideous manifestation of unbridled biodiversity, and it is perhaps only the tip of the iceberg in human-spider relations. It bodes poorly for the likelihood of any kind of planetary arachnid-human diplomatic summit.

Next morning, with Vicodin pulsing through my veins keeping me calm, I found an oxygen tank in my shop and fashioned a fitting that would deliver oxygen to the left side of my nose. I figured that if hydrogen peroxide would be good, pure oxygen at high pressure would be better, and maybe it would absorb deeper into the receding skin. Every hour while I was awake, I spent fifteen minutes with this tube tightly pressed to my nose. The lesion there didn't enlarge like the rest, which were getting bigger every day. If I had owned the right sort of fittings, I would have put oxygen on all of them. I was practicing a form of triage.

The third day, it hurt to move, and I took Vicodin just to get up in the morning. The spiders were out of it by now because I had blasted the house with spider bombs. I slept better, but I was still stuck to the sheets in the morning. The wounds don't die with the spiders.

Shelly Hendler is my physician and my good friend. When I need a doctor, I have Shelly. Even in the middle of the night. And I trust him. When I first described the spider bites, he wanted to know how I had concluded it was the brown recluse. I told him that the *Merck Manual* described the lesions precisely. He was convinced when I told him a day later that I had killed every insect in the house with a spray bomb and that there were several brown recluses among the bodies.

Shelly checked the medical books, some friends, and the

Internet. There seemed to be nothing good that you could do for the brown recluse's bite, short of surgery. Keep it clean, put peroxide on it, hope it goes away. Don't count on it. Call a surgeon.

Shelly diagnosed me over the phone. He was 600 miles away in San Diego. He said it sounded like the wounds were infected with bacteria and that I should take penicillin. I said, "Shelly, penicillin doesn't work against toxins."

"Well, it sounds like bacteria."

"I'm sure bacteria are there feasting on the wound, but they aren't the problem." I was right, but also I was wrong. The problem was that my immune system couldn't defend me against bacteria. The spiders had fixed that with their venom. Penicillin could have killed the bacteria directly without need of the immune system, but I wasn't thinking clearly.

Shelly asked me if I needed more Vicodin.

"It hurts like shit."

"I'll call it in. You sure you won't take penicillin?"

"No. I think it might weaken me."

Shelly was unsettled.

I took Vicodin for the next eleven days. The wounds got worse. Every morning I would check them with a ruler, and they were growing. No sign of any healing.

The pain got worse, especially the one on my right elbow. The pictures over the Internet of people with brown recluse bites were disgusting. Some had warnings: *The material you are about to see is graphic and disturbing!* Holes in skin, pus dripping. I was scared but not scared enough to return to San Diego and get medical intervention. Vicodin in large doses interferes with pain and with judgment.

On day eleven, I called Shelly to request more analgesia. I told him that it was getting more painful. He was concerned, and having decided that I was no longer a reliable witness of my own problem, he wanted to fly up the next day without telling me. He was willing to do anything that he and I thought reasonable. He's not opposed to pain relief, having had some pain in his life, but he really wanted me on penicillin. He said, "I'm going to prescribe some dicloxacillin. Pick it up when you go into Ukiah for the oxycodone."

Nothing in the medical literature said anything about penicillin for *Loxosceles* bites. Surgery seemed to be the only answer. In my case, with ten bites, surgery would be a massacre. I would lose the function of one knee, maybe an elbow, and it would leave deep scars everywhere. Big hunks of my skin and my muscles would be laying there in the stainless steel bucket by the time it was over.

I decided that I had nothing to lose. I picked up the prescription in Ukiah and started with half a gram of dicloxacillin at about three in the afternoon. I took another half gram at six, another at nine, and then I went to sleep with the aid of the oxycodone. I woke up at three in the morning and amazingly, I was not sticking to the sheets! My wounds didn't hurt. I went into the bathroom to the mirror. The wounds, although still round, were turning into skateboard abrasions. The most beautiful scabs I had ever seen were forming on my arms and legs.

My bout of necrotic arachnidism was over. Shelly Hendler had discovered the cure for brown recluse spider wounds.

We didn't scientifically prove it because we haven't tested it again and again. We're busy, and I, for one, don't care to expose myself to the brown recluse again to see if it works. But I would definitely recommend dicloxacillin for brown recluse spider wounds. Unless you are allergic to penicillin, it won't hurt you. I kept taking it for about a week until the scabs were dry and falling off. It worked on all ten wounds.

As for the brown recluse spider, I say kill the bastards any time you see them. They have six eyes and eight legs. I think that's too many of each. Biodiversity be damned. I'd be glad to step on the last survivor of the *Loxosceles* genus personally.

13

NO ALIENS ALLOWED

SOME PEOPLE HAVE experiences that are so strange, they attribute them to alien intervention of some kind. Close encounters of the first kind, second kind, third kind, etc., as though alien intervention would always fall into certain categories. I had one of those experiences myself. To say it was aliens is to assume a lot. But to say it was weird is to understate it. It was extraordinarily weird.

In 1975 I bought some property about ten miles inland along the Navarro River in Mendocino County, California. Rather than call it "The Firs" or "Sunshine Hill" or "The Mullis Place," I called it "The Institute for Further Study." Somewhat later, I renamed part of it "Fire and the Rose Automatic Tree Farm." Tree farms were favored by the IRS as a form of investment, and I was planting trees, they were automatically growing, and my pond was an earthen reservoir from which the water to grow trees was being taken. So I became a tree farmer. I still am. I never had the heart to cut the trees down, so I didn't show a profit within five years, and I can't claim it any more. America is stronger because of my trees, and I'm proud of them, even though my business failed. I don't think that had anything to do with the fact that one night I got spirited away by mysterious beings. I'm almost sure they weren't IRS people.

I was living in Berkeley at the time, and I'd drive up to my property on Friday nights. One night in 1985 I got there just around midnight. I had driven up alone, and I had passed the functional sobriety test—I had made it through the mountains.

I turned on the kitchen lights, put my bags of groceries on the floor, and grabbed a heavy, black flashlight. I was headed to the john, which was about fifty feet west of the cabin, down a hill. Some people thought it was a little eerie at night, but I didn't—I liked the night. I liked sitting in the dark on the custom carved redwood seat. I liked the sound of owls in the valley. But that night, I never made it to the seat.

The path down to the john heads west and then takes a sharp turn to the north after a few earthen steps. Then it runs level for about twenty feet. I walked down the steps, turned right, and then at the far end of the path, under a fir tree, there was something glowing. I pointed my flashlight at it anyhow. It only made it whiter where the beam landed. It seemed to be a raccoon. I wasn't frightened. Later, I wondered if it could have been a hologram, projected from God knows where.

The raccoon spoke. "Good evening, doctor," it said. I said something back, I don't remember what, probably, "Hello."

The next thing I remember, it was early in the morning. I was walking along a road uphill from my house. What went through my head as I walked down toward my house was, "What the hell am I doing here?" I had no memory of the night before. I thought maybe I had passed out and spent the night outside. But nights are damp in the summer in Mendocino, and my clothes were dry, and they weren't dirty.

The lights in the cabin were dim. I quickly turned off the switch. Miles from Pacific Gas and Electric, I had my own solar panels and a couple of batteries under the house. It was adequate but not deluxe. I was always careful about the lights. The grocery bags were still on the floor, and I started putting them away. The freshly squeezed orange juice from the Safeway in Healdsburg was no longer cold. My memory of the night before was slowly returning. I recalled that I had headed to the john with my nice new black flashlight. Where the hell was that?

All of a sudden, it came back to me. The talking, glowing raccoon! Had that happened? It was as clear a memory as my early morning brain allowed. Yes. I remembered the little bastard and his courteous greeting. I remembered his little shifty black eyes. I remembered the way my flashlight had looked on his already glowing face. Where was my flashlight?

I walked to the john right away. I wasn't afraid of finding something scary. I wanted that fucking raccoon to be there. It wasn't—and neither was the flashlight. I had a feeling nothing would be there. I had a feeling I was going to feel empty, frustrated, and confused, and I was.

I was also sleepy. I went back to the house and climbed in bed and slept for several hours. When I awoke, the experience took on a much sharper reality. I checked once more for the flashlight. I still couldn't find it, even after I expanded my search around the property. All the facts—my dry clothes, the house lights on all night, my flashlight—were things that I couldn't deny, and yet I didn't panic. I couldn't call anyone because I didn't have phone service. The whole thing seemed totally perplexing.

I searched again for my new flashlight, to no avail. I decided to go about my regular business of the day. There seemed to be no way to investigate it further. The most unusual thing about it was that it did not bother me as much as it should have. I decided to clean out a pipe.

There is a spring in the most beautiful part of my woods. Water from that spring normally flows through a pipe and feeds a pond. I'd noticed the week before that the pipe needed cleaning, so late in the afternoon, I headed into the woods with a few tools. The woods are about 200 yards from the house across an open meadow.

Just inside the shade of the trees, I began to panic. I turned around and walked as rapidly as I could toward the daylight. I didn't run or look over my shoulder—I walked fast. I didn't want anything to know that I was panicking. When I got well out into the open, I turned and looked back into the woods. "What the hell am I doing?" I had no idea. But I wouldn't go back in there. Each time I looked in that direction, I felt more certain that I wouldn't go back in there.

Whatever happened to me last night must have happened there, in those woods. I remembered that the road I had been walking on that morning when I came to my senses came toward my house from that direction. I got the hell home and didn't go back. I didn't tell anybody about it.

Six months later I walked in those woods with my two boys. They were five and eight, and with them I somehow felt more comfortable. We spent a little time there. I unclogged the pipe. But I didn't return alone for some time, and I still didn't talk about it.

It was weird having a part of my property where I didn't feel comfortable. I was in Mendocino a lot of the time by myself. Why had I suddenly developed an irrational fear of a place I'd always enjoyed?

A year or two passed. One Saturday night when I was up by myself for the weekend, I decided to take matters in my own hands with some therapy. I had wanted to go clear out that goddamned pipe again during the afternoon. I had assembled the tools. But I had resisted going over there. Instead of going dancing at the Rose Bud saloon that night, I would do some psychotherapy.

I had purchased another black metal flashlight to replace the one I had lost. I taped it to the barrel of an AR-15. This is a weapon not everyone has, thank God. I was excused from the Vietnam War, so the first time I saw one was when a friend brought it to Mendocino. When I saw it, it looked like a toy from Mattel. Ron assured me that it wasn't. The clip held about twenty shells, and it would fire a new one every time you pulled the trigger. They were legal, and I felt good having it around—my place was isolated and had no phone.

With a five-D-cell flashlight strapped to the barrel of the AR-15 with black tape, I felt like John Wayne. I walked out to the woods. I stood outside the first trees and yelled into the dark. "This is my property, and I'm coming in. Anything moves—I'll shoot it. If it doesn't move, I may shoot it anyhow. I'm pissed off." I was yelling really loud. "Get out of my woods. Now. If you can't move, scream. Maybe I'll have mercy. Maybe not. Get the fuck out of my woods." John Wayne would not have said "fuck," but times have changed.

I felt like that kind of screaming would at least clear out anybody who was innocently there. It was also part of the therapy. My D-cells cut a clean beam into the darkest part of the forest. There was a giant old hollowed bay laurel growing right out of a little waterfall full of ferns. I was fifty feet away from it. I loved it, but it had become the focus of my fears. John Wayne, at my side, wearing the same kind of hat I had donned for the occasion, said, "Let 'em have it, kid." I opened up with the AR-15 and riddled the area of the laurel. "Blow 'em to hell, kid!"

I emptied one clip and loaded another, walking around and screaming and firing at anything that looked dark. I didn't shoot up in the air—I'm not antisocial.

The psychotherapy worked. I hoped that I hadn't shot any holes in my water line. I walked out of the woods knowing that in the morning, without the AR-15 or the hat, I could come back. And I did.

Some time later I was in a bookstore in La Jolla. I noticed a book on display by Whitley Strieber called *Communion*. On the cover was a drawing that captured my attention. An oval-shaped head with large inky eyes staring straight ahead.

I bought the book and immediately began reading it. It was Strieber's personal account of being abducted by aliens. He wrote of waking up in his cabin in the woods of New York State and seeing an owl staring at him. He spoke to the owl, then two beings, who looked like the figure on the cover of the book, appeared in his doorway and escorted him out of the house. He wrote that he had smelled burning cinnamon and smoldering cheese around them, so I burned some myself to see if that might excite a memory, but it didn't.

While I was reading this book my daughter, Louise, called from Portland. "Dad, there's a book I want you to read. It's called *Communion*."

"I'm reading it right now."

She began to tell me what had happened to her in Mendocino. She had arrived at the house with her fiancé late one
night. And just like me, she had wandered down the hill. She
was gone for three hours. Her fiancé had spent the time franti-
cally searching for her everywhere, calling her name, but she
was nowhere to be found.

The first thing she remembered was walking the same road
on which I'd found myself, hearing her fiancé calling her
name. She had no idea where she had been.

When she saw the book, she had experienced the same sort
of vague recognition as I had. After she finished telling me her
story, I told her about my experience. It was the first time I'd
told anyone. I asked her about talking raccoons that glowed in
the dark.

"I don't remember anything," she said.

Strange things have happened in Mendocino. My closest
neighbor there, Alex Champion, who was in graduate school
with me at Berkeley, thinks that there are a lot of mysteries in
the valley. He doesn't think they cause problems, as long as
we recognize that we are the Real Things and that we are in
charge. We only have to take command, he says. I think John
Wayne would agree.

I wouldn't try to publish a scientific paper about these
things, because I can't do any experiments. I can't make glowing raccoons appear. I can't buy them from a scientific supply
house to study. I can't cause myself to be lost again for several
hours. But I don't deny what happened. It's what science calls
anecdotal, because it only happened in a way that you can't
reproduce. But it happened.

14

THE 10,000TH DAY

ONE SUMMER MY elder son Christopher had three jobs. He was a hardworking boy. His younger brother Jeremy, who had zero jobs that summer, decided that Chris was a loser and a dweeb. I like Jeremy in spite of his superior attitude toward everybody in the solar system because he knows how to buy groceries for a whole week, how to transform them into splendid meals, and how to shame Christopher into cleaning up. To give Jeremy something to do over his vacation, I calculated how many days old he was and suggested he keep track of it on a calendar. After a moment of consideration, Jeremy announced that I was a loser and, sadly, also a dweeb.

Jeremy doesn't appreciate numbers.

I do. Some numbers are intriguing. 0, 1, 1, 2, 3, 5, 8, 13, 21, 34 . . . for instance. See where that goes? It continues on forever. By adding the last two terms together, you get the next one. It's called the Fibonacci series, and if that isn't enough excitement for you, then you divide the last term by the one right before it and as the numbers get larger, you get a closer and closer approximation to something called Phi. Try it. 5 divided by 3 is 1.667. 8 divided by 5 is 1.600. 13 divided by 8 is 1.625. And on and on it goes, until you have computed the ratio of the long side of the Parthenon to the short side of the same building. It's not the way the Greeks did it. They thought of it as a ratio of the long piece of a five-pointed star's side divided by the short piece, if you can catch the Greek drift. If you can't, just draw a five-pointed star and start measuring the line segments, and you will quickly understand. It only matters because I am trying to make an argument here that I am not a dweeb. Jeremy is a dweeb. Numbers are as fun as groceries.

THERE IS A 10,000th day in your life. It creeps up on you and nobody sends you a card. It happens about three months after your 27th birthday. Most people fail to take the day off. Too bad. I wasn't about to go into the lab that day. I went to a nude beach near Santa Cruz. I put my blanket down on about 10 billion grains of sand and let a thousand waves wash over my toes while I watched eleven naked women play in the surf.

You missed your 10,000th day? Don't worry. Maybe you can catch your 20,000th day. You will be fifty-four and about nine months old then. You have to figure it out from your own day of birth and take all the leap years into account. And don't forget about this: on the 13th day of your 57th year, you will be 500,000 hours old. During those 500,000 hours, your heart will have beat about 2.25 billion times. You will have breathed in and out about 300 million times, which just happens to be the number of meters that a beam of light travels in one second, and also the number of 1992 U.S. dollars that Hoffmann-

La Roche paid Cetus Corporation for the patent rights to my PCR invention without bothering to even send me a card celebrating my 17,520th day, or make it 301 and send me the change. Screw Cetus and the Swiss! My father's side of the family came from Flums. And Flums is still reluctantly under the Swiss flag. But Flums is on the Liechtenstein side. The Hoffmann's live with their dark brown insect associates on the Basel side in that part of Switzerland that neither the Germans nor the French coveted enough to capture. A part of Switzerland, mind you, where they don't even celebrate the Fourth of July.

Because the Earth is not a clock made in Switzerland, it doesn't have any gears with a certain number of teeth, so it doesn't turn exactly 365 times around on its polar axis every time it completes a full circle around the sun. We call its rotations "days" because of the sun coming up at the beginning of everyone of them, and we call revolutions around the sun "years" because that's how long it takes and we have one birthday in each one, and Hoffmann-La Roche continues to fail to send me even a birthday card on mine. Never mind. What this business about the lack of teeth and cogs means is that years are not evenly divided into days. The number of days in a year is not a number like the number of eggs in a dozen. It is somewhere around 365.2425 . . . and some more, days per year, which believe it or not was actually figured out in 1582 without computers, or the Internet, by astronomers working for, of all people, the Church—Pope Gregory XIII, presiding.

What this means is that every four years, we have to add a day to the number of days we say are in a year to let the Earth catch up with our count. That's why we have leap year, and that would be all she wrote if 0.2425 . . . and some more, was equal to 0.2500, but it isn't. So we don't actually have leap years every four years. We have them almost every four years on the years that are evenly divisible by four—like 1996. But in those years, divisible evenly by four, that are also evenly divisible by one hundred, we don't add an extra day to February and we don't have a leap year. For example, 1900 was not a leap year, 1904 was. But it gets more complicated than that, and it gives you a little more respect for the guys who were counting days for the Pope in the sixteenth century—some 300 years, mind you, before the invention of the latex condom. If the year is evenly divisible by 400, like the year 2000 is going to be, we do have an extra day in February and a leap year. The only exception is years that are evenly divisible by 4000. That's the rules. These years are numbered, of course, from the birth of Holy Jesus. Was that in AD 0 or would that have been AD 1? And if it were AD 0, would it have been a leap year? At least Jesus didn't have any trouble remembering how old he was, plus or minus one, of course.

These rules are called the Gregorian calendar, and they have been helpful in organizing events in history. The work done, to establish them, was paid for by the Church, the purpose of which was to prevent Easter from ever happening on the Fourth of July.

All that being as it may well be, you can count the number of days you have seen, and I suggest you do. It's hard for people who travel a lot, and especially for people like Story Musgrave, who spend a lot of time in orbit. But you can put a mark on a stone that you carry in your briefcase every midnight, or tell your computer that whenever t = 00:00:00, then $N = N + 1$. You can know how many suns have come up while you have been alive. Every now and then it will be cause for taking the day off.

But on those days that you have to go to work, what can you do about the continual passage of time, and the fact that the morning mail delivered to your "in" basket always brings you some new variety of grief? Caught in the giant clockwork, you wish that it were Friday afternoon, at one minute to five, and everyone was tidying up their desk. Couldn't you just live in that moment forever? John Kenneth Galbraith, in a book called *The Affluent Society*, written before the Age of CNN, suggested that the reason James Watt harnessed steam in 1765 was to allow us all, eventually, to live in that final 1 billion nanoseconds before five o'clock on Friday afternoon. "Ahhhh! Weekends," he thought, as he strolled across a well-mown lawn in Scotland and realized that if the steam were condensed by a piece of metal in contact with the boiler, a steam engine would work so well that no one else would have to.

Perhaps I was more enthusiastic about this than either Watt or Galbraith, but in 1982 I discovered that the Industrial Revolution had truly arrived. I put a simple laboratory robot on my desk at Cetus and programmed it to relieve me of the scourge of passing days. On entering my office in the A.M., I could activate a toggle switch on the robot's base. It would elegantly swing its arm around, doing a little preliminary dance that I had choreographed for it. It was a lovely instrument. Just a rotating base with a single arm, ending in what they referred to prosaically in the handbook as the "gripper." A graceful hand with two padded fingers. The hand would slide into my "In" box and carefully grip the mail. It would slide out, just as gracefully, and holding itself perfectly balanced over the trash can, open the grippers. The grippers would then close delicately, and the arm would swing around and downward. It would pause, for a moment, to savor the occasion, then press the closed gripper onto the off-side of the toggle switch. The robot and I would have finished our day, living out the dream of the Industrial Revolution.

I have become lazier. Counting one's days can be exhausting. This coming winter, I plan to retract my feelers from the whole process, temporarily, while I take a long nap. Bears do it, and not a single forest ranger complains. Next winter will be my fifty-fourth. Fifty-four is nine times six and six is two times three, and three to the second power is nine. That makes it a certainty. I'll sleep next winter from December until the end of February, except for Superbowl Sunday—unless the cumulative score in the playoffs is evenly divisible by 400.

Jeremy may come up to Mendocino for Christmas and find me sleeping, instead of celebrating the 17.51411 millionth hour since the birth of Christ. When I get up in February, I will begin a correspondence with him by e-mail about which of us is a dweeb.

15

I AM A CAPRICORN

I BEGAN TO think about astrology in the mid-1960s after three strangers had correctly classified me as a Capricorn. The probability of that happening by chance is 1 out of 1,728.

The first was Emma, a ten-year-old neighbor in Atlanta, where I was a student at Georgia Tech. I was walking up the steps with my groceries when she proclaimed, "You're a Capricorn, aren't you?" I stopped in my tracks. How did she know? I asked her how a Capricorn acts.

She replied, "Like you."

If Emma was just guessing, it was a good guess. There are twelve signs that your sun can be in when you are born. When people say that you are a Pisces or a Capricorn without being any more specific, they are saying that the sun was in that part of the sky called Pisces or Capricorn when you were born. So they have a one out of twelve chance of being right.

At your birth, the Moon, Venus, Mars, and the rest of the planets are also in some particular part of the sky, but those things move around in their own pattern and only people who are more conversant with astrology than Emma would concern themselves with them. When Emma told me I was a Capricorn, I didn't know anything about astrology.

When you look at all the planets and include the sun and moon, their relative positions define a shape at the moment of birth. That shape represents the Native, the person being charted, in an overall way. Several of the planets might be arrayed in a really striking pattern. Or they may not be. In mine, the planets are spread all around but there are these two ominous sets of three planets referred to as T-squares. In my first wife's chart, there are three planets in a perfect equilateral triangle called a Grand Trine. T-squares predict that the Native will have a hell of a time getting his shit together and be late handing in manuscripts and may do a short time in jail or even worse. A Grand Trine means that the Native will be born with a silver spoon. She might be lazy, but she knows what she has and that's exactly what she needs. That made sense to me from what I knew about us. When I did our daughter Louise's chart, I found that she was a perfect blend of our charts. Louise had the shape of an Aquarian Kite. That's like a Grand Trine mated to a T-square, with her Ascendant headed into her mother's Aquarian sun sign. Out of phase with the kite, she shared a Capricorn sun with me. Totally weird, I thought.

I knew Louise would reflect us genetically—but astrologically?

The next time somebody came at me out of the blue with my sun sign was three years later in Berkeley. I was at a party talking to some woman and she stopped in mid-sentence. "You're a Capricorn. I know it."

How did she know?

She said it was the way I was waving my hands when I talked. And the way that I held onto the countertop when I was not waving them. I was also leaning forward, then backing off.

In terms of the number of people who had told me my sign, and the number of people who had been right, that was two for two. They both could have been guessing. It's one in twelve. Two for two on a one in twelve is one in a hundred and forty-four.

Being a scientist, the important thing to me was the long odds. When something unusual happens, a scientist worth his thick horn-rimmed glasses and shoddy clothes gets moving. I went back to the astrology books, drew up a few more charts for my friends, and decided that in order to save myself a lot of calculation time and trips to the library, I would write a computer program to do that for me. That turned out to be difficult. Isaac Newton had written down the rules for how things move around each other due to gravity. It was fairly easy, knowing the starting points for two things like the Earth around the sun, to predict where the Earth would be a hundred or even a thousand years later. A computer program could easily do the math. But the problem with the solar system is that there is not just one planet. There are too many planets. Each of them is affected, not just by the sun, which it dutifully orbits, but by each of the other planets. The big ones like Jupiter and Saturn have the greatest effect, but even the little ones make their little perturbations every time they make a close encounter, and after a hundred years things get fairly complicated. Naval Observatory astronomers had been writing programs for years trying to simulate the movements of the planets and they were pretty accurate, but they were still working on it. There were reasons other than astrology for this work by the Navy. Things like navigation and satellites and trying to drop a missile into Red Square.

One night about a month after that party in Berkeley, I was

camping by the Navarro River in Mendocino County. People were walking all around from fire to fire and some guy stood outside of our circle listening to me tell a story. When I was done, he stepped into the light and announced that I was a Capricorn. He turned around, and I called to him.

"How do you know?"

He turned. "Because of the way you come on, really strong and then back off. You act like one." He left haughtily, a swagger in his step, like a goddamned Scorpio.

Three for three of one in twelve—1 out of 1,728. That's the probability of three consecutive people independently announcing your sign correctly.

I was convinced that it was not a matter of chance. Those people were observing my behavior and making a reasonable estimate of my sun sign. If people can really do that from a little bit of information, then astrology is worth investigating.

One little experiment I did was by accident. I had my chart done by a shop in La Jolla that sent your birth date, time, and place, to a company in L.A. that used a computer to do the calculations and then to select a number of paragraphs about you from a huge number that they had about everybody. It was what you would call a computerized expert system. Most of the things that the fifty-page document said about me were correct. But some of them were entirely wrong. It turned out that the ones that were wrong were derived from my rising sign.

The rising sign in a chart is sensitive to the time of birth more than anything else in the chart. It is the part of the sky that is coming up over the eastern horizon at the time and place that you are born. It changes every minute.

The computer assumed that someone would not really know what *actual* time he was born if he was born during World War

II in America. We had an extra hour of daylight savings time. In 1944, when I was born, if your birth certificate said that you were born at 1:53 PM in December, you were really born at 12:53 PM. I knew that when I filled out the form. I put in the right time and called it EST rather than EWT. The computer figured I didn't know what I was writing and corrected EWT to EST. The result was that I got a horoscope that was an hour off. My moon was misplaced just a half degree to the west. Against the backdrop of the stars, the moon moves slowly toward the east, not to be confused with its apparent movement to the west caused by the earth's rotation. But my rising sign was way off. It was Taurus instead of Aries.

Being educated in these things, I was more entertained than damaged. God forbid I had been dependent exclusively on that computer to tell me about myself.

To be an Aries rising and to mistakenly think that you are a Taurus rising could cause you to conclude that you were fucking up. A Taurus rising feels himself to have physical substance, he takes care of things like a farmer, he doesn't depend on others a lot because he knows they can't be trusted. His humor is ironic if at all, and he is thoroughly fixed. He is a mountain. He does not pray for he knows that nothing changes. But he believes.

An Aries rising feels his oats, but not his substance. He does new things. He is alone and so he originates. He has a conscience because everything that happens is his fault, but he can behave excessively since no one else is there. He dares. He prays. But he does not believe.

I knew that a mistake had been made when I read the paragraphs that were based on my rising sign.

The rest of the printout was correct. I wondered whether

someone familiar with me, but not with the fault in this rendering of my horoscope, could determine which of the various pronouncements was wrong.

I gave the printout to a really good friend who didn't know anything about astrology. I asked him to go over the two hundred or so items about me and put an *x* beside any that he thought did not apply to me. He did. Almost exclusively he marked those items that were derived from the bogus rising sign.

I had copied the printout so nobody could see his *x*'s because I am a scientist. I tried to find more people who were willing to look at my horoscope seriously. I found two. They also put *x*'s most often by the paragraphs that had to do with my misplaced rising sign.

I explained the error to the people with the computer and they redid my horoscope with the correct time. The new one fit. Once more I asked friends to mark passages that didn't apply to me. There were fewer *x*'s and they weren't concentrated on items from the rising sign.

From all this I can conclude a number of things. A horoscope that accurately reflects your personality can be cast by a computer if you give it the correct birth data, and at least three of my friends know me at least as well as a computer program. It was entertaining and a pretty cheap experiment. Little girls, people at parties, and voices out of the darkness by the Navarro River can tell you what month you were born in.

We consider ourselves to be sophisticated, intelligent, modern people. Our psychologists and sociologists consider astrology to be nonsense. Academic departments concerned with human behavior consider astrology to be a confusing distraction, with no serious value to their pursuits. And it's not that they've never heard of it. They've noticed that every daily paper in the world has a column devoted to it and that lots of humans pay attention to it. The reason they don't pay attention to it is that it would embarrass them in front of their colleagues. There's no proven body of facts in the social sciences that says human behavior does not contain elements that are related to planetary positions at the time of birth. Instead, there's a broad and arrogant understanding among social science professionals that folklore, like astrology, is for simpletons. Without doing any simple experiments to test some of the tenets of astrology, it has been completely ignored by psychologists in the last two centuries.

Most of them are under the false impression that it is nonscientific and not a fit subject for their serious study. They are dead wrong. Whether or not the present-day practitioners of astrology are using scientific methods has no direct bearing on whether the body of knowledge they employ is true and valid. To have dismissed it without any experimental evaluation as insubstantial drivel from the masses says a lot about the fact that the present-day mental health practitioners have their heads firmly inserted in their asses and generally need more help than they provide.

We know little about ancient astrology besides the fact that as long as five thousand years ago civilizations ranging from Babylonia to China independently looked to the heavens for help in understanding life on Earth. In the seventeenth century, when men like Galileo, Kepler, and Newton were laying the foundations of astronomy, they were also concerned with the astrological significance of the observations they were recording and learning how to predict. Somewhere along the line, though, the precision that they could bring to the act of measurement and mathematical prediction must have outweighed the usefulness of the thoughts they could bring to bear on the rather more vague concepts that astrology required. Men who stay up all night looking through long black tubes, recording numbers with four or five digits, and inventing calculus don't necessarily know a whole lot about human beings, and they aren't likely to take an interest in the complex interactions between people and the stars. They've got enough to worry about just trying to figure out why the orbit of Mars is elliptic instead of circular.

So astronomy separated itself from astrology. But not because one worked and the other didn't. No one did extensive empirical testing of astrological facts and concluded that nothing useful could be predicted from any of it. Astronomers just preferred to stick to the cyclical movements of planets rather than the cyclical movements of people.

They specialized in the numbers. And astronomy is a rich and interesting field because of it. Behold the nice pictures of things very far away that the Hubble telescope sends back.

But astrology is still here and it could be a valuable tool for understanding human beings if serious students of behavior would lower themselves to examine it. Are there any serious students of behavior? Medical researchers have for a long time recognized that folk remedies often work. Ethnobotanists examine the healing use of herbs by primitive people, who don't know what molecules are, but when the herb works, it works, and therefore it gets incorporated into scientific medicine. If nobody knows how it works, somebody finds out. Folklore is a rich source of new information. But you don't hear about modern psychologists out mining the world of folklore for new concepts. You don't hear about it because it's not done.

They're stuck with a loose set of theories of learning and behavior that completely ignore a vast area of human understanding that begins with the premise that all men are definitely not created equal. They are divided into a complex array of different types that can be at least sorted out, if not partially understood, by looking at the positions of the planets in the sky at the site and time of their birth. Preposterous, but it is true, and it is scientifically accessible. Furthermore, these various types of people are affected differentially by the continued movement and rearrangements of those same planets for the rest of their lives. They come in and out of cyclical bursts of creativity, periods of deep depression, warm fulfilling experiences, horrible losses, and on and on.

How can somebody call himself a student of human behavior and hang out a shingle offering to help humans solve their problems without at least studying astrology? How could an institution of higher learning grant someone a Ph.D. in psychology without requiring at least a few courses in astrology? If psychologists were doing okay, that is, if they had a good track record for freeing their patients from the pain that they pay good money to sort out and be relieved of, then I could see why the good head doctors could thumb their noses at the folklore of astrology, but *nobody* would be so demented as to imagine for a moment that when you go to a shrink you get anything resembling good mental health. If you are lucky in your choice of psychologist, maybe you won't do yourself in this year, but no one expects a human in chronic emotional pain to get a miracle cure. In other words, psychology is practiced by a bunch of well-paid incompetents. They can't fix a broken heart.

They ought to be looking around for some new theories. Freud, Jung, Maslow—they were cool, fun to read maybe— but we're still neurotic, and some of us still jump off bridges. Astrology by itself is not the answer to all our problems any more than herbs from the Amazon witch doctor, but it's a shame to waste such a vast and ancient resource because of the simple fact that our modern witch doctors are too frozen in their attitudes to take a look around.

I don't go to shrinks. Would you take your car to a mechanic who refused to acknowledge the existence of separate makes and models?

Astrology also contains a deep mystery or two that should whet the appetite of any curious student of "what's going on in the universe." How the hell does my brain have any way of knowing about the relative position of the planets before I learned how to use the *Nautical Almanac*? It must somehow be in touch with these things either directly or indirectly since it seems to be affected by them. And the "how" of that should be as interesting to a physiologist as to a sociologist, or a psychiatrist, even a physicist. The fact that it is correlated with these things can be easily established by observing the nonrandom distribution of birthdays among various professions.

A recent scientific study of the distribution of medical students in birth months discovered that a lot of medical students were born in late June. They postulated that it was because the sun was up earlier and so there was more light for them right away and they could be outside and therefore would get interested in biology. Well, that was bullshit. It's the same in Australia, and the sun is not up early in June down in the antipodes. Successful applicants to medical school do not come equally from each month. They cluster around Gemini–Cancer in both hemispheres. More biochemists are born in Sagittarius. Lawyers have their own distribution, and some people claim reasonably that lawyers hatch from eggs and eat their own young—not enough obviously—so they have their own separate problems. Sociology has so far turned a blind eye to these things. It could be that's one of the reasons sociology is so boring and such a worthless science. It's pedantic and uninformed.

I was born at 17:58 Greenwich Mean Time on December 28, 1944 in Lenoir, North Carolina. You can find out more about me from that than you can from reading this book.

16

THE AGE OF
NUTRITIONAL OBSESSION

NANCY POINTED OUT to me this evening, in a book she bought by a certified nutritionist, that margarine should be avoided. The author goes so far as to suggest that it is a villain.

I don't like margarine either. I never eat it, and Nancy never serves it. I like butter, but I spent fifteen minutes on the Internet because I wanted to know why the author would say something like that. The Internet is like a library in your home. Even late at night, I can find out whether claims are supported by established facts or whether they were just made up for effect.

It doesn't take a lot of education to check things out. All it takes is access to resources and a minor distrust of everyone else on the planet and a feeling that they may be trying to put something over on you. At the end of her book, the author had a long list of references supposedly supporting her conclusions. While she listed the references, she failed to specify exactly what books supported which arguments. She didn't make it clear where she had learned what she had claimed she had learned, making it difficult to check up on her. Scientific method takes issue with this kind of callous disregard for the impersonal nature of knowledge.

I searched the Internet for "trans fats." I found twenty-eight references. One was relevant. It referred to a study which concluded that, compared to saturated fats alone, margarine might cause a minor change in the ratio between so-called good cholesterol and bad cholesterol. This information should not be reason to call margarine "villainous."

Cholesterol has been shown to be an important indicator of cardiovascular disease, or not, depending on which reports you believe. But there are a lot of clinics that do cholesterol profiles and make a lot of money at it. People believe in the numbers they generate, and they try all kinds of things to raise their level of good cholesterol and lower the level of bad cholesterol. Now, if doing this makes people happy, they should keep doing it. But it doesn't make any sense. No one has any hard evidence that all this stuff about good cholesterol and bad cholesterol makes any difference.

This is what we know about cholesterol. It makes up a considerable percentage of the membranes surrounding every single one of our cells. We make cholesterol ourselves, and we control the amount of it that we make. Cholesterol synthesis in humans is connected to the synthesis of hormones like androgens and estrogens, which are connected to all of our sexual functions. Chemists think of them as cholesterol derivatives. Cholesterol is not some horrible thing that chickens put into their eggs, it's something our bodies need, otherwise we wouldn't be making it. If there was something wrong with it, our bodies would have learned how to make something else to replace it.

The human digestive system transforms everything we eat, including cholesterol, into unrecognizable bits of matter before turning it into us. The process begins in the stomach, where the foods we eat are subjected to hydrochloric acid and some horrendous catalysts that begin breaking it down. In our intestines, we change the scene. Now without the acid of the stomach, we subtly slip sharp enzymatic blades into the molecules that are left and sever their structures into units of universal biologic currency. Only these little pieces of universal life matter are allowed into our bloodstream. Through the portal vein they are admitted to our liver. When our liver gets through with the meal it sends it out to the rest of our body in a form that is so dissected down to the lifeless elements of earth that no cell anywhere in our body except our annoying brain can know exactly what we had for dinner.

There are a few things, however, that we cannot afford to break down into their constituent elements because we have lost the ability to re-form them. These are the compounds that we call vitamins. It is our need for vitamins that got us into our present nutritional obsession.

Vitamins are little bits of organic matter that most earthly organisms can assemble out of the basic elements that our liver generates. Sometime during the long, pretelevision, days of evolution, our cells forgot how to make things like vitamin C out of the stuff coming up out of our liver. It was an accident. It had to do with random mutations in our DNA that, at the time, didn't matter. We were already getting plenty of vitamin C coming up out of our liver. It came from our food and was not broken down in our stomach, intestines, or liver. It emerged into the bloodstream intact, readily available to all of our cells. We lost the ability to make it ourselves, without becoming extinct, because we didn't need to make it. And more than that, we reproduced faster than our associates who had not lost the ability to make it.

There is a very important rule in evolution: *Don't trouble yourself with details that do not matter for survival*. Evolution has to deal with intense competition—like a race car driver always on the last lap, wishing he had just enough gas to get him to the finish line. Whoever can do something more efficiently survives. Losing the ability to make something that you already have in plentiful supply is efficient. In the long run it limits your options but in the last lap, streamlined is better than being burdened with useless tools.

We also lost the ability to make a number of other molecules that we need. Like vitamin C, they were readily available in the leaves and fruits we were eating and did not get processed into something unrecognizable in the digestive system or the liver. Now, a couple of million years later, we have assigned the letters A, B, C, D, and E to them.

One day, about 2 million years ago, we decided to exercise one of our options. We came down out of the trees and stopped eating only leaves and fruit. It was a good move because Africa was drying up into more grass and fewer trees. The trees were dark and more and more filled with things that were feeling the pressures of a shrinking habitat.

While we were up in the trees, we had noticed that there were awkward-looking, grass-eating animals on the edge of the forest. They would later be called ungulates and still look awkward today. We came down from the trees in little gangs, chased those ungulates around until they got tired, bashed them with sticks, and ate them. Eventually we discovered that the meat tasted a lot better when cooked, but unknown to anyone on the planet at the time, the fire also destroyed vitamin C. So those of our ancestors who ate only well-done steaks and no green leaves died of scurvy, while those of us who retained a taste for salads and fruit survived.

The first humans who dropped the salad habit, in an organized fashion, were the sailors. The hearty mates who set sail from Europe in the fifteenth century ran out of salad about the same time that they started missing their girlfriends on shore. They began to succumb to scurvy after three months, and it was ugly. The first sign was that their teeth and fingernails started bleeding. It looked contagious because somebody came down with it first and then sailors all over the boat started showing symptoms. The first cases were thrown overboard in a futile effort to save the rest.

One day, a lucky scurvy victim was dropped off on an island with citrus trees. He started eating oranges and miraculously recovered. He made his way back to England and, on arriving, announced to the Admiralty that scurvy was not infectious but rather was caused by a nutritional deprivation. Ships were then fitted out with large barrels of limes and the British sailors were thereafter referred to as limeys. It made an incredible difference in maritime activity, and it started a nutritional mania that is now reaching ludicrous heights.

For some reason, people today have accepted the claim that a normal diet can't possibly satisfy all of our nutritional needs and that many of the things that we eat are very bad for us. There are "nutritionists" who present themselves as experts on eating. They are not biochemists, nor are they incredible chefs. They tell us that just eating a normal human diet is not enough, and we can buy their books to learn the right way to be healthy.

There are several molecules that we do need and that we have lost the ability to make because they were plentiful in our food at some time during our evolution. There are only a few of these, but they are heavily represented in our present diet. If you isolated someone from normal foods and fed them only things without these molecules, they would get sick.

But how did we get from there, the concept that the esthetic existence of our bodies is determined by certain essentials in our food, to the concept that the esthetic appearance of our bodies is shaped by a very careful balance of carbohydrates, proteins, unsaturated fatty acids, no milk shakes, no saturated fatty acids, no really fat fatty acids, omega-3-fatty acids, no chocolate, no eggs, no pizza, no hamburger, undulating fatty acids, spine-tingling fatty acids, and fingernails chewed to the quick out of the fear of an improper diet?

Some people eat too much; some people eat too little. Nothing else about diet really matters. Check out the reality of things and it will make you feel better. Logically established facts allow you to sleep better at night, which is essential, even in the presence of creatures howling in the dark and nutritionists who write diet books. Throw the books out and shut the door. You'll sleep better.

17

BETTER LIVING
THROUGH CHEMISTRY

I STARTED USING drugs when I was a child. I got them from my mother. She started me off on barbiturates. Remember The Thing? People trapped in the Arctic with a monster you couldn't see, my brother Robert on the floor behind the seats in front of us asking me what was happening. How about Them? Giant ants in Nevada that moved to New York. Night would come. Robert would be sound asleep. He'd only *heard* about giant ants in the sewers with venom dripping off their two-foot stingers. I could still see them. Mom would give me phenobarbital. Phenobarb was considered by my mom and her physicians to be a reasonable way to overcome a sleepless night. It has long since been replaced with more expensive things like Valium.

Once in a while, Mom would give me codeine for pain.

When I had a cold she'd buy me a Benzedrine inhaler. There was a small piece of cotton inside the inhaler that was saturated with amphetamine free base. It was a little off-white plastic tube that cost thirty-nine cents and fit easily into my pocket. I could sit directly in front of Mrs. Coleman, my first-grade teacher, snorting speed. It relieved a stuffy nose, and when a cold was trying to put you down, it picked you right up.

If first graders tried to get away with that kind of behavior today, they'd never see the bright lights of second grade.

When I had a cough, or diarrhea, I got paregoric, a solution of 10 percent opium in alcohol. Opium comes from a handsome poppy. It cures diarrhea. It also cures coughs, and when you feel bad, it makes you feel better.

No one warned my mother that she was doing something bad. She was just giving me the medicines that good mothers had always given their children, drugs that could be purchased at the local pharmacy. Eventually all of these wonderful things became illegal. It was hard for me to understand exactly why a drug like paregoric, one of the most useful drugs in all the pharmacopoeia, should suddenly be considered dangerous and outlawed. Nobody seemed to mind, however, even my mother didn't complain.

Kansas became the final state to outlaw paregoric, in 1976. I was living in Kansas in 1976, working with Richard Zakheim, a pediatric cardiologist. I was headed to Mexico for a vacation and Dick wrote me a prescription for paregoric just in case. It may have been the last bottle.

The summer following my freshman year at Georgia Tech in 1962 I was working with Al Montgomery in our lab, trying to purify a solution of para-phenyl benzoic acid in benzene. It was worth about forty dollars a gram and I had made almost fifty grams. As I was boiling the benzene on a hot plate, it burst into flames and blew flaming benzene all over my hand. I wrapped my shirt around it and Al and I raced twenty miles in rush hour traffic to the hospital. We drove the old blue '55 Chevy up on sidewalks, went through lights, did things that would have attracted a police escort any other time. No luck.

We raced into the emergency entrance of the Baptist Hospital. I waited with my hand in a stainless steel pan of warm sterile solution for ninety minutes while they tried to find a doctor. Many doctors walked blandly by the door. Finally a surgeon arrived and gave me something for the pain. One minute I felt like I was holding burning coals in my hand—then the morphine came down my arm. It was the most pleasant feeling I'd ever had. I could still feel a sensation, but it wasn't pain. More like a pleasant itch.

At Georgia Tech some of my friends used speed. They were part of the academic tradition of staying awake all night to cram for exams. No one considered speed "drugs." The Sigma Chi fraternity house bought them from the housemother—they were available by prescription for weight control. Our housemother was conveniently overweight. Nobody called her a pusher. It was the 1960s.

At Georgia Tech, I had a wife and a little girl. I had short hair and I studied all the time. My senior year I made perfect grades. I studied physics and math and chemistry to the point where I would never have to study them again. And all I knew about drugs was what I read in magazines like *Time* and *Life*. I learned that marijuana was a dangerous addictive drug and that I should stay away from it. On the other hand, I learned that LSD was a miracle that just might enable scientists to understand the workings of the brain, could be the cure for alcoholism, and, just incidentally, might prevent World War III. Psychiatrists were prescribing it for their patients. In 1966 LSD had not yet been made illegal. Respected, well known people were admitting that they had experimented with LSD. The Luce family, the publishers of *Time* and *Life*, were so intrigued by the scientific potential of LSD that they funded the research of Harvard professor Timothy Leary.

A person who loved playing with chemicals as much as I did just couldn't help but be intrigued by LSD. The concept that there existed chemicals with the ability to transform the mind, to open up new windows of perception, fascinated me. I considered myself to be a serious scientist. At the time it was still all very scholarly and still legal. There was no tawdry aura over it. People weren't blaming their kids' problems on it yet. Hippies had just started to differentiate themselves from beatniks and the difference seemed to be fewer years and more hair on the hippies. And they stayed in college.

In 1966 I wanted to try LSD. My wife, Richards, helped me pack up the Impala, we put our daughter Louise in the back seat, and we drove to Berkeley for graduate school. It was the first time I'd been to California and it surprised me. I had not expected that the trees would be different. I didn't know that the Pacific Ocean was always cold. I didn't expect San Francisco to be foggy in the summer. I thought there would be naked girls. I certainly didn't know that I would be changed so profoundly.

I didn't want to take LSD alone. I had learned that from magazines. The first week of class I became friendly with the only guy in my class with long hair, Brad. I figured he would have LSD. Brad was smart. He appreciated the fact that I could calculate how long it would take the moon to fall to the Earth. He had graduated from Oberlin College, where they knew it was possible to do such a calculation but they wouldn't be so crass as to actually learn how.

Brad had experimented with psychedelic drugs and agreed

to guide me through my first trip. He suggested that before I took LSD, I should smoke some marijuana because it might give me some idea of how my consciousness would be changed. Marijuana scared me, I told him. Everything I'd read about it said that it was a bad drug, an addictive drug—one toke and you're a slave for life.

He persuaded me to smoke a "joint," as he called it. Within moments my fear disappeared. I was laughing. Brad and I talked about wise things for hours. At some point, Brad left. I looked at Richards, my wife, with new eyes. She was the same Richards, but not to me. I grabbed her in a primitive way, rolled her onto our enhanced bed, and felt the surging power of bliss.

A week later I said, "Brad's going to come over tonight. I'm taking acid." Richards said she would make a nice dinner.

During dinner, Brad gave me what was called a double-domed 1000 microgram Owsley. He had bought it for five dollars. It was soon to become illegal. I didn't finish dinner. I started laughing. I got up from the table and realized, on the way to the couch, that everything I knew was based on a false premise. I fell down through the couch into another world.

Brad put *Mysterious Mountain* by Hovhaness on the stereo and kept playing it over and over. It was the perfect background for my journey. I watched somebody else's beliefs become irrelevant. Who was that Kary Mullis character? That Georgia Tech boy. I wasn't afraid. I wasn't anything. I noticed that time did not extend smoothly—that it was punctuated by moments—and I fell down into a crack between two moments and was gone.

My body lay on the couch for almost four hours. I felt like

I was everywhere. I was thrilled. I'd been trapped in my own experiences—now I was free. The world was filled with incredibly tiny spaces where no one could find me or care what I was doing. I was alone. My mind could see itself.

Brad had given me 1000 micrograms because he wanted me to have a thorough experience. I think he said "blow your ass away." With 100 micrograms you feel a little weird, you might hallucinate, and you can go dancing, but you know you're on acid. You're aware that you're having a trip and the things that you see are hallucinations. You know that you should not respond to them. When you take 1000 micrograms of LSD, you don't know you've taken anything. It just feels like that's the way it is. You might suddenly find yourself sitting on a building in Egypt three thousand years ago, watching boats on the Nile.

After four hours Brad told me we were going to take a ride in the car. I didn't know what a car was. We got inside this thing and it started moving and I started to panic. I didn't want to be in a car. I didn't like movement, I just wanted to find a quiet place. Eventually we stopped in Tilden Park by a fountain. I got some water. It was cold and fluid but it wasn't the water I knew. It left trails and it was alive. I didn't know Brad, I didn't know my wife. When they got me back in the car I understood I was inside a vehicle. I knew it had a key that made it work, but I didn't want it to. I was sitting in the back seat, and we started down Marin Avenue, which drops 800 feet in four blocks. Berkeley was below and I was dizzy. I reached over from the back seat and pulled out the key. Brad took back the key, told me to behave, and drove home.

About five o'clock in the morning I began to come back to

earth. The most amazing aspect of the entire experience was that I landed back in the middle of my normal life. It was so sweet to hear the birds, to see the sun come up, to watch my little girl wake up and start playing. I appreciated my life in a way I never had before.

On the following Monday I went to school. I remember sitting on a bench, waiting for a class to begin, thinking, "That was the most incredible thing I've ever done."

I wrote a long letter to my mother. I often wrote to my mother to tell her what I was thinking about. As I was writing the letter, I began to realize that for the first time in my life, there were some things that I might not be able to explain to her. But I tried.

My mother responded by sending me an article she'd torn from the *Reader's Digest*. It said that taking LSD was bad for your brain and will cause flashbacks for the rest of your life. She entreated me not to do it anymore. I wrote back that it was too late. It had already changed me.

I wanted to understand what had happened. How could 1000 micrograms—one thousandth of a gram—of some chemical cause my entire fucking sensorium to undergo such incredible changes? What mechanisms inside my brain were being so drastically affected? What did these chemicals do to my visuals? I wanted to know how it worked. I wanted to know more about neurochemistry.

Berkeley had a classic biochemistry department, meaning it consisted of professors who specialized in the chemical mechanisms underlying all life. They didn't know much about mammals, besides their wives and students, and they weren't interested in neurotransmitters. I was on my own. I knew that my brain was behind my eyes. I learned that no one knew very much about how it functions. We knew which parts of the brain controlled certain things, but we didn't know how or why. It seemed pretty obvious to me that neuroactive drugs might help us find out. These chemicals caused a really interesting interaction between psychology, biochemistry, and anatomy, but we didn't know why. There was good reason to expect that we might learn something about mental illnesses, which might be caused by an imbalance in the chemistry of the brain.

A s we learned very quickly, LSD was not the only mind-altering chemical. When it became illegal, we started synthesizing other chemical compounds. It usually took the government about two years from the time the formula for a new psychoactive compound was published to make it illegal. Numerous derivatives of methoxylated amphetamines were created, for example, and every one of them had a different effect on the brain.

I was very careful to make only legal compounds. Other people were not. And the authorities were serious about this business. People were going to jail for chemistry. Once, someone in the chemistry department got arrested. I was working in Joe Neilands's lab at that time, and he responded to a bust by dropping a copy of the Berkeley *Gazette* on my desk. "They're getting pretty close to home, don't you think?" he remarked. "If there's anything in the freezer that shouldn't be there, maybe now would be a good time to clean it out." Then he walked away. Joe treated his students as adults, but he didn't want to visit me in jail.

Drug laws don't have much to do with science or health.

Opium was made illegal in California because Chinese dock workers in San Francisco were taking jobs away from Irish dock workers who preferred to be drunk than opiated. Opium dens were raided and Chinese workers were arrested. Conveniently, they couldn't report for work in the morning. They moved north.

Marijuana was declared illegal after the end of Prohibition in 1938 because the opium/alcohol cops needed something to police or they'd lose their jobs. To gain public support, marijuana was depicted as a dangerous drug that caused black and Mexican men to lust after white women. It wasn't the drug. Black men and Mexican men didn't suddenly develop a need for white women; white men suddenly developed a need, after 1938, for jobs. Alcohol was back in; marijuana was shortly going to be out. People who wanted to be into prohibition would now prohibit marijuana. The same people, and maybe their children, would be happy to make a living prohibiting LSD.

LSD somehow got connected with the anti–Vietnam War movement. Drugs had to be the reason that the youth of America had long hair, wore beads, enjoyed sex, and didn't think it was a good idea to go to a foreign country and kill the locals. Psychedelic drugs were made illegal.

The one serious effect it had, besides putting a lot of people in overcrowded jails, was to bring to an end serious research by people who knew what to look for. The only scientists permitted to work with psychoactive chemicals now were people who never used them and knew nothing about them. For the first time, science reference books were censored. Standard chemical reference books like the *Dictionary of Organic Chemicals* eliminated all mention of LSD and methamphetamine. How dare they censor reference books? It was as if an entire class of chemicals no longer existed. It was getting darker in America.

Every drug experience was unique.

While always interesting, it was not always fun. I sometimes visited very dark places. During the year after I left my wife and daughter, my pain was magnified every time I took acid. I had hurt other people, and I felt my guilt. I thought I was the ugliest person in the world.

There was one trip from which I thought I would never come back. I thought I had destroyed my physical brain. My friend Eric, with whom I often did psychedelic drugs, was a strategic air command pilot. In the air, he was responsible for one of the keys required to arm the nuclear bombs. If war started, he had partial responsibility for dropping them. One day he realized he couldn't, and wouldn't, do it. They gave him an honorable discharge. Psychiatric problems—he wouldn't help blow up the world.

One weekend while he was still on active duty, he was staying with Richards and me. I had synthesized diethyltryptamine. Not much was known about it. I expected the effect to be similar to dimethyltryptamine, but lasting longer. I weighed out what would be a reasonable dose, but I made an error. I must have had a premonition because I told Eric that I would take it first, we'd wait a half hour until it took effect, and then if it was all right, he would join me. I took ten times the amount I had intended. Within a few minutes something was terribly wrong. The last sane words I said were, "Don't take it, Eric."

It was too much. The fire roared out of the fireplace. I was no longer in the room. I was somewhere lying on a gurney, being wheeled down a hospital corridor. Not on Earth. My friends were playing a joke on me. They were sending me to

Earth to be born. To them, it was like sending me to a scary movie. They didn't know I was going to spend a lifetime on this planet far, far away. It was just a joke. And then I realized that this had already happened. I am here. I'm stuck. I don't know how to get home. I wanted it to stop, but I couldn't speak.

I woke up in the living room and saw a snake coming out of the fireplace. I found a piece of wood and started beating the snake. It was Eric's clarinet. He was unhappy about the clarinet, but he and Richards were far more worried about me.

I woke up the next morning huddled under my desk. Everything was gray. I couldn't remember who I was, what I did, what I liked. I was terrified and sad. I looked out the window and saw children playing in the yard. One of them was mine, but I didn't know which one. Richards woke up. She told me she was my wife, but I didn't remember her. Nothing in my house was familiar. I thought I loved books and music, but I couldn't remember which books or what kind of music.

I had annihilated my personality. I had no preferences. I didn't recognize my body. I wasn't physically uncomfortable. I could walk around. I could eat. I had no friendships, no love, no humor. Eric and I had often gone camping along the Navarro River. He thought that might be a good place for me to be. That evening as we sat by a campfire he read me a poem he said I liked. "Do you remember what a poem is?" he asked. I remembered parts of it, but only from a distance.

In the morning my memory slowly began to come back. In another day it was back completely. Whole and undamaged. I was functioning normally, and my personality was back. I felt that I had been to some very important place. I now knew what it felt like to be psychotic, to be meaningless. But it sure as hell hadn't been fun being lost.

18

CASE NOT CLOSED

WHEN I FIRST heard in 1984 that Luc Montagnier of France's Pasteur Institute and Robert Gallo of America's National Institutes of Health had independently discovered that the retrovirus HIV—Human Immunodeficiency Virus—caused AIDS, I accepted it as just another scientific fact. It was a little out of my field of biochemistry, and these men were specialists in retroviruses.

Four years later I was working as a consultant at Specialty Labs in Santa Monica. Specialty was trying to develop a means of using PCR to detect retroviruses in the thousands of blood donations received per day by the Red Cross. I was writing a report on our progress for the project sponsor, and I began by stating, "HIV is the probable cause of AIDS."

I asked a virologist at Specialty where I could find the reference for HIV being the cause of AIDS.

"You don't need a reference," he told me. "Everybody knows it."

"I'd like to quote a reference." I felt a little funny about not knowing the source of such an important discovery. Everyone else seemed to.

"Why don't you cite the CDC report?" he suggested, giving me a copy of the Centers for Disease Control's periodic report

on morbidity and mortality. I read it. It wasn't a scientific article. It simply said that an organism had been identified—it
did not say how. It requested that doctors report any patients
showing certain symptoms and test them for antibodies to this
organism. The report did not identify the original scientific
work, but that didn't surprise me. It was intended for physicians, who didn't need to know the source of the information.
Physicians assumed that if the CDC was convinced, there
must exist real proof somewhere that HIV was the cause of
AIDS.

A proper scientific reference is usually a published article
in a reliable scientific magazine. These days the magazines
are slick glossy paper with pictures on the front and lots of
advertisements, a lot of editorial material by people who are
professional journalists, and a few pictures of girls selling you
things you might want to buy for your lab. The advertisers are
the companies who make things for scientists to buy and the
companies who make drugs for doctors to sell. There are no
major journals without advertisements. Therefore, there are no
major journals without corporate connections.

Scientists submit the articles in order to report their work.
Preparing articles describing their work and having them published is crucial to a scientist's career, and without articles in
major journals, they will lose their rank. The articles may not
be submitted until experiments supporting the conclusions
drawn are finished and analyzed. In primary journals, every
single experimental detail has to be there either directly or by
reference, so that somebody else could repeat exactly what
was done and find out whether it comes out the same way in
their hands. If it doesn't, somebody will report that, and the

conflict eventually has to be resolved so that when we go on from here, we know where "here" is. The most reliable primary journals are refereed. After you send in your article, the editors send copies of it to several of your colleagues for review. They become the referees. The editors are paid for their work on the journal; the referees are not. But what they do gives them power, which most of them like.

I did computer searches. Neither Montagnier, Gallo, nor anyone else had published papers describing experiments which led to the conclusion that HIV probably caused AIDS. I read the papers in *Science* for which they had become well known as the AIDS doctors, but all they had said there was that they had found evidence of a past infection by something which was probably HIV in some AIDS patients. They found antibodies. Antibodies to viruses had always been considered evidence of past disease, not present disease. Antibodies signaled that the virus had been defeated. The patient had saved himself. There was no indication in these papers that this virus caused a disease. They didn't show that everybody with the antibodies had the disease. In fact, they found some healthy people with antibodies.

If Montagnier and Gallo hadn't really found this evidence, why was their work published, and why had they been fighting so hard to get credit for the discovery? There had been an international incident wherein Robert Gallo of the NIH had claimed that a sample of HIV which had been sent to him by Luc Montagnier of the Pasteur Institute in Paris had not grown in Gallo's lab. Other samples collected by Gallo and his collaborators, from potential AIDS patients, had grown. Gallo had patented the AIDS test based on these samples, and the

Pasteur Institute had sued. The Pasteur eventually won, but back in 1989 it was a standoff and they were sharing the profits.

I was hesitant to write "HIV is the probable cause of AIDS" until I found published evidence that would support it. Mine was the most minimal statement possible. In my grant request I wasn't trying to say that it absolutely did cause AIDS, I was just trying to say that it was likely to cause it for some known reasons. Tens of thousands of scientists and researchers were spending billions of dollars a year doing research based on this idea. The reason had to be there somewhere, otherwise these people would not have allowed their research to settle into one narrow channel of investigation.

I lectured about PCR at innumerable meetings. Always there were people there talking about HIV. I asked them how it was that we knew that HIV was the cause of AIDS. Everyone said something. Everyone had the answer at home in the office in some drawer. They all knew and they would send me the papers as soon as they got back. But I never got any papers. Nobody ever sent me the news about how AIDS was caused by HIV.

I finally had the opportunity to ask Dr. Montagnier about the reference when he lectured in San Diego at the grand opening of the UCSD AIDS Research Center, which is still run by Bob Gallo's former consort, Dr. Flossie Wong-Staal. This would be the last time I would ask my question without showing anger. In response Dr. Montagnier suggested, "Why don't you reference the CDC report?"

"I read it," I said, "That doesn't really address the issue of whether or not HIV is the probable cause of AIDS, does it?"

He agreed with me. It was damned irritating. If Montagnier didn't know the answer, who the hell did?

ONE NIGHT, I was driving from Berkeley to La Jolla and I heard an interview on National Public Radio with Peter Duesberg, a prominent virologist at Berkeley. I finally understood why I was having so much trouble finding the references that linked HIV to AIDS. There weren't any, Duesberg said. No one had ever proven that HIV causes AIDS. The interview lasted about an hour. I pulled over so as not to miss any of it.

I had known of Peter when I was a graduate student at Berkeley. He had been described as a truly brilliant scientist who had mapped a particular mutation to a single nucleotide in what was to become known eventually as an oncogene. In the 1960s, that was a real feat. Peter went on to develop the theory that oncogenes might be introduced by viruses into humans and cause cancer. The idea caught on and became a serious theoretical driving force behind the research that was funded under the unfortunate name "War on Cancer." Peter was named California Scientist of the Year.

Not satisfied resting on his laurels, Peter torched them. He found flaws in his own theory and announced to his surprised colleagues who were working on demonstrating it that it was highly unlikely. If they wanted to cure cancer, their research should be directed elsewhere. Whether it was because they were more interested in curing their own poverty than cancer, or that they just couldn't come to grips with their mistake, they continued to work fruitlessly on the viral oncogene hypothesis for ten years. And they didn't seem to notice the irony: the more frustrated they got, the more they chastised Peter Duesberg for questioning his own theory and their folly. Most of them had not really learned much of what I call science. They had been trained to obtain grants from the government, hire people to do research, and write papers that usually ended with the notion that further research should be done along these same lines—preferably by them and paid for by someone else. One of them was Bob Gallo.

Gallo had been a friend of Peter's. They had worked in the same department at the National Cancer Institute. Of the thousands of scientists who had worked fruitlessly to assign a causal role in cancer to a virus, Bob was the only one who had been overzealous enough to announce that he had. No one paid any attention because all he had demonstrated was an anecdotal and very weak correlation between antibodies to a harmless retrovirus, which he called HTLV I, and an unusual type of cancer found mainly on two of the southern islands of Japan.

In spite of his lack of luster as a scientist, Gallo had worked his way up in the power structure. Peter Duesberg, despite his brilliance, worked his way down. By the time AIDS came along, it was Bob Gallo whom Margaret Heckler approached when President Reagan decided that enough homosexuals picketing the White House was enough. Margaret was the Secretary of Health, Education, and Welfare and thereby the top dog at NIH. Bob Gallo had a sample of a virus that Luc Montagnier had found in the lymph node of a gay decorator in Paris with AIDS. Montagnier had sent it to Gallo for evaluation, and Bob had appropriated it in the pursuit of his own career.

Margaret called a press conference and introduced Dr. Robert Gallo, who suavely pulled off his wraparound sunglasses and announced to the world press, "Gentlemen, we have found the cause of AIDS!" And that was it. Gallo and Heckler predicted that a vaccine and a cure would be available within a couple years. That was 1984.

All the old virus hunters from the National Cancer Institute put new signs on their doors and became AIDS researchers. Reagan sent up about a billion dollars just for starters, and suddenly everybody who could claim to be any kind of medical scientist and who hadn't had anything much to do lately was fully employed. They still are.

It was named Human Immunodeficiency Virus by an international committee in an attempt to settle the ownership dispute between Gallo and Montagnier, who had given it different names. To call it HIV was a short-sighted mistake that preempted any thought of investigation into the causal relationship between Acquired Immune Deficiency Syndrome and the Human Immunodeficiency Virus.

Duesberg pointed out wisely from the sidelines in the *Proceedings of the National Academy of Science* that there was no good evidence implicating the new virus. He was ignored. Editors rejected his manuscripts and committees of his colleagues began to question his need for having his research funds continued. Finally, in what must rank as one of the great acts of arrogant disregard for scientific propriety, a committee including Flossie Wong-Staal, who was feuding openly with Duesberg, voted not to renew Peter's Distinguished Investigator Award. He was cut off from research funds. Thus disarmed, he was less of a threat to the growing AIDS establishment. He would not be invited back to speak at meetings of his former colleagues.

WE LIVE WITH an uncountable number of retroviruses. They're everywhere—and they probably have been here as long as the human race. We have them in our genome. We get some of them from our mothers in the form of new viruses—infectious viral particles that can move from mother to fetus. We get others from both parents along with our genes. We have resident sequences in our genome that are retroviral. That means that we can and do make our own retroviral particles some of the time. Some of them may look like HIV. No one has shown that they've ever killed anyone before.

There's got to be a purpose for them; a sizable fraction of our genome is comprised of human endogenous retroviral sequences. There are those who claim that we carry useless DNA, but they're wrong. If there is something in our genes, there's a reason for it. We don't let things grow on us. I have tried to put irrelevant gene sequences into things as simple as bacteria. If it doesn't serve some purpose, the bacteria get rid of it right away. I assume that my body is at least as smart as bacteria when it comes to things like DNA.

HIV didn't suddenly pop out of the rain forest or Haiti. It just popped into Bob Gallo's hands at a time when he needed a new career. It has been here all along. Once you stop looking for it only on the streets of big cities, you notice that it is thinly distributed everywhere.

If HIV has been here all along and it can be passed from mother to child, wouldn't it make sense to test for the antibodies in the mothers of anyone who is positive to HIV, especially if that individual is not showing any signs of disease?

Picture a kid in the heartland of America. His lifelong goal has been to join the Air Force when he graduates and become a jet pilot. He's never used drugs and he's had the same sweet girlfriend, whom he plans to marry, all through high school. Unbeknownst to him, or anyone else, he also has antibodies to HIV, which he inherited from his mother, who is still alive, when he was in her womb. He's a healthy kid, it doesn't bother him in any way, but when he is routinely tested for HIV by the Air Force, his hopes and dreams are destroyed. Not only is he barred from the Air Force, but he has a death sentence over his head.

The CDC has defined AIDS as one of more than thirty diseases accompanied by a positive result on a test that detects antibodies to HIV. But those same diseases are not defined as AIDS cases when the antibodies are not detected. If an HIV-positive woman develops uterine cancer, for example, she is considered to have AIDS. If she is not HIV-positive, she simply has uterine cancer. An HIV-positive man with tuberculosis has AIDS; if he tests negative he simply has tuberculosis. If he lives in Kenya or Colombia, where the test for HIV antibodies is too expensive, he is simply presumed to have the antibodies and therefore AIDS, and therefore he can be treated in the World Health Organization's clinic. It's the only medical help available in some places. And it's free, because the countries that support the WHO are worried about AIDS. From the point of view of spreading medical facilities into areas where poor people live, AIDS has been a boon. We don't poison them with AZT like we do our own people because it's too expensive. We supply dressings for the machete cut on their left knee and call it AIDS.

The CDC continues to add new diseases to the grand AIDS definition. The CDC has virtually doctored the books to make it appear as if the disease continues to spread. In 1993, for example, the CDC enormously broadened its AIDS definition. This was happily accepted by county health authorities, who receive $2,500 from the feds per year under the Ryan White Act for every reported AIDS case.

In 1634 Galileo was sentenced to house arrest for the last eight years of his life for writing that the Earth is not the center of the universe but rather moves around the sun. Because he insisted that scientific statements should not be a matter of religious faith, he was accused of heresy. Years from now, people looking back at us will find our acceptance of the HIV theory of AIDS as silly as we find the leaders who excommunicated Galileo. Science as it is practiced today in the world is largely not science at all. What people call science is probably very similar to what was called science in 1634. Galileo was told to recant his beliefs or be excommunicated. People who refuse to accept the commandments of the AIDS establishment are basically told the same thing. "If you don't accept what we say, you're out."

It has been disappointing that so many scientists have absolutely refused to examine the available evidence in a neutral, dispassionate way. Several respected scientific journals have refused to print a statement issued by the Group for the Scientific Reappraisal of the HIV/AIDS Hypothesis simply requesting "a thorough reappraisal of the existing evidence for and against this hypothesis."

I spoke publicly about this issue for the first time at a meeting of the American Association for Clinical Chemists in San Diego. I knew I would be among friends there. It was a small part of a much longer speech—at most I spoke for fifteen minutes about AIDS. I told the audience how my inability to find a simple reference had sparked my curiosity.

The more I learned, the more outspoken I became. As a responsible scientist convinced that people were being killed by useless drugs, I could not remain silent.

The responses I received from my colleagues ranged from moderate acceptance to outright venom. When I was invited to speak about PCR at the European Federation of Clinical Investigation in Toledo, Spain, I told them that I would like to speak about HIV and AIDS instead. I don't think they understood exactly what they were getting into when they agreed. Halfway through my speech, the president of the society cut me off. He suggested I answer some questions from the audience. I thought it was incredibly rude and totally out of line that he cut me off, but what the hell, I would answer questions. He opened the floor to questions and then decided that he would ask the first one. Did I understand that I was being irresponsible? That people who listened to me might stop using condoms? I replied that fairly reliable statistics from the CDC showed that in the United States, at least, the number of reported cases of every known venereal disease was increasing, meaning people were not using condoms, while using the initial definition of AIDS, the number of reported cases of AIDS was decreasing. So, no. I didn't understand that I was being irresponsible. He decided that that was enough questions and ended the meeting abruptly.

Whenever I speak on this issue the question always comes up, "If HIV isn't the cause of AIDS, then what is?" The answer to that is that I don't know the answer to that any more than Gallo or Montagnier. Knowing that there is no evidence that HIV causes it does not make me an authority on what does. It is indisputable that if an individual has extremely close contacts with a lot of people, the number of infectious organisms that this individual's immune system is going to have to deal with will be high. If a person has three hundred sexual contacts a year—with people who themselves are each having three hundred contacts a year—that's ninety thousand times more opportunity for infections than a person involved in an exclusive relationship.

Think of the immune system as a camel. If the camel is overloaded, it collapses. In the 1970s we had a significant number of highly mobile, promiscuous men sharing bodily fluids and fast life styles and drugs. It was probable that a metropolitan homosexual would be exposed to damn near every infectious organism that has lived on humans. In fact, if you had to devise a strategy to collect every infectious agent on the planet, you would build bathhouses and encourage very gregarious people to populate them. The immune system will fight, but the numbers will wear it down.

The scientific issue gets tangled up with morality. What I'm describing has nothing at all to do with morality. This is not "God's wrath" or any other absurdity. A segment of our society was experimenting with a life style and it didn't work. They got sick. Another segment of our pluralistic society, call them doctor/scientist refugees from the failed War on Cancer, or just call them professional jackals, discovered that it did work. It worked for them. They are still making payments on their new BMWs out of your pocket.

19

HAVE SLIDES
WILL STAY HOME

I WAS INVITED by the Glaxo Pharmaceutical Company to speak at a conference. They sent me a letter in December of 1993 asking me to be the November 1994 symposium banquet speaker. If that time was not convenient for me, they wanted me to speak at the November 1995 banquet. Dr. John Partridge, who was the director of the Chemical Development Division, had not met me personally but had heard about a lecture I had given in 1991 at the Gordon Research Conference that, in his words, was "the most highly praised lecture that I have ever heard about from my academic and industrial colleagues."

He was looking for "particularly articulate scientists who bridge the biochemical and medical disciplines and routinely engage in 'out of the box' thinking."

Well, that certainly was me.

Dr. Partridge wrote that he would be pleased to pay all my travel and accommodations, as well as an honorarium of $1,500.

I thought this sounded all right, but I figured Glaxo could pay me a little more. What made this invitation particularly interesting to me was the fact that Glaxo was the largest drug company in the world, and one of their profitable drugs was the cellular poison being used against AIDS, AZT. It kills cells like a cancer chemotherapeutic does. It keeps them from reproducing by preventing them from making new DNA. It also kills HIV. In cancer, there is a rationale at least for using them, although I personally would never use chemotherapeutics on myself, cancer or not. But here's the way the explanation goes.

I think it stinks of an old therapy they used to use against syphilis, arsenic. The syphilis was surely going to kill you, the arsenic might kill you, but maybe it would kill the syphilis first and you would live to fraternize again. The use of poisonous chemotherapeutics in cancer follows the same line. The cancer is surely going to kill you. The chemotherapeutic surely will also, but maybe it will kill the cancer cells before it kills you. It's a gamble. We will give you almost enough to kill you and hope it's sufficient to kill the cancer. I wouldn't go for it myself. I don't need to take drugs that make my hair fall out. But what the hell, if somebody wants to take this kind of gamble, it does have a sort of logic to it. Nothing fun. Nothing you would do for a headache. But it's a chance somebody might want to take when the alternative is to die too young to watch their kids grow up. And some people do recover from cancer even after they have taken chemotherapeutics.

In the case of AIDS, the same strategy took a diabolic turn. AIDS might kill you, AZT might also. It will surely make you sick. It will prevent the proliferation of any rapidly growing cells in your body including the CD-4 immune cells that your doctor thinks you need now more than anything. It may kill the HIV. It kills it in petri dishes. But that may not cure you. The damage to you may have already been done, whatever it is.

The complete absence of all HIV from your body, even if it is accomplished, may not cure you of AIDS. No one has ever recovered from AIDS, even though they have recovered from HIV. And we are not going to give it to you in a limited dose as we do in the case of cancer chemotherapy, where we are gambling that although we are hurting you, we are hurting the cancer more and maybe you will survive longer. Here we are not gambling. No one has ever recovered from AIDS. We cannot expect that you might recover. We are going to ask you to swallow this poison until you die.

About a half a million people went for it. No one has been cured. Most of them are dead. The ones that are not are also taking another drug now, a protease inhibitor. Who knows what it will do? The manufacturers didn't know when they started selling it. The FDA didn't require them to show that it would cure AIDS and not kill the patient any more than they required them to show that about AZT. They only required that a surrogate goal be met. A surrogate goal means that something that we think may be related to the disease in question may be improved by the drug, like the level of CD-4 cells, whatever the fuck they are. It's a way to get around the notion that a drug ought to be effective in curing the disease that it is sold for before it can be sold. The surrogate goal bullshit is an indication that our FDA no longer serves our needs. Or at least it does not serve our needs unless we own stock in the pharmaceutical industry and don't give a shit about health care.

I was interested in giving a seminar about things like this to the scientists assembled in North Carolina by Glaxo, formerly Burroughs Wellcome, and by the University of North Carolina in the name of Frontiers in Chemistry and Medicine. I was thinking that this technique of killing people with a drug that was going to kill them in a way hardly distinguishable from the disease they were already dying from, just faster, was really out there on the edge of the frontiers of medicine. In previous interviews and seminars I had said that I thought AZT was not only useless against AIDS, but in fact it was poisoning people. There were large-scale medical studies done in Europe, called the Concorde Study, that indicated just this. AZT was worthless against AIDS and harmful even to healthy people. This conclusion was reached despite the fact that the study was heavily funded by Glaxo. I wondered if these people knew how I felt about their product when they issued the invitation.

I notified Dr. Partridge that I was pleased to accept if they would raise the ante a little. On January 26, 1994, I received a letter from M. Ross Johnson, the Vice President of the Division of Chemistry. They were very happy that I had accepted and wrote that they would send me first-class airfare for two, accommodation expenses, and an honorarium of $3,000. In closing, he asked me for the title of my banquet presentation.

So far, so good. I responded as requested, explaining that I intended to speak to this audience about a subject that should be of tremendous concern to the entire scientific community. I would speak about the fact that there is no scientific evidence that HIV is the probable cause of AIDS and that I believed people taking the drug AZT were being poisoned.

On October 14, 1994, a month before the meeting, I received another letter from Glaxo—this time from Gardiner F. H. Smith. No title. He was sincerely regretting having to inform me that they could no longer accommodate my presentation. He said that they would send me a check for $1,000 to compensate me for any inconvenience.

I responded with the following letter:

Dear Mr. Johnson:

Enclosed please find a copy of a fairly uninformative letter from a Mr. Gardiner Smith, with whom I have not been in contact or correspondence previously.

As you know, my overall schedule is compact and very difficult to rearrange on short notice. I have declined, as a result of my commitment to Glaxo, income from other potential engagements. With Mr. Smith, I sincerely regret that your company has been forced into the "changing of the structuring," whatever that means to Mr. Smith, of "the above-referenced event."

Unfortunately, I have made arrangements to attend to several nonprofit institutional functions in the Southeast in connection with this trip, appearances which I will not cancel. Therefore, your company's reluctance, as related perfunctorily by Mr. Smith, to abide by the terms of your (previous) correspondence represents a considerable loss of income as well as an unanticipated expense to me personally.

Mr. Smith's unexplained offer of $1000 compensation for my "time and trouble" adds a bit of mystery here as to who Mr. Smith is and what he must misconceive to be the value of my time and trouble.

I do not understand what Mr. Smith is exactly apologizing for in his letter, but I will be kindly expecting immediately, with or without an explanation from some more cordial and informed representative of Glaxo, a check for $6048.00.

For Mr. Smith's information, round-trip airfare between San Diego and Raleigh-Durham first class for two is $3048. Addition of our agreed-on honorarium of $3000 results in the above figure.

One more thing you might consider, Dr. Johnson. A number of attendees at your meeting will likely have something to say to me about my failure to appear. You should be careful to explain there publicly precisely why Mr. Smith felt the need to inform me that your company has taken the liberty of "restructuring" in such a way as to be unable to "accommodate" my presentation. I am not in the habit of canceling public appearances at such short notice and would not care to gain such a reputation on your account. I hope you understand that this is not for me or for Glaxo, a trivial matter.

Cordially,

Dr. Kary B. Mullis

On November 30, 1994, I received another letter from Mr. Smith. It was quite brief, saying that he had received a copy of my letter to Dr. Johnson. Enclosed was a check from Glaxo in the amount of $6,048.

This was the most money I had ever made specifically for not doing something. And it occurred to me that, with my growing reputation for creating controversy, there might be many groups or individuals who did not want to hear me speak. Certainly that was their right, but if people did not want to hear ideas that would make them uncomfortable, they ought to be willing to pay not to hear them. With that thought in mind, I drafted the following offer:

HAVE SLIDES WILL STAY HOME

Dr. Kary B. Mullis wants to talk to you and your associates, your friends, your sons and daughters. Is there anything you can do about it?

YES . . . BUT YOU MUST ACT NOW . . .
SPECIAL OFFER

Dr. Mullis won the Nobel Prize in Chemistry in 1993 and promptly launched a worldwide lecture tour. Universities, research institutes, conventions, high schools, *businesses, community groups,* he even addressed "Connect"—a joint project of UCSD and the San Diego biotech industry—*right on the beach in front of his very own apartment, which has been described in the national press as "rented rooms filled with his tools of seduction."*

He is usually invited to lecture on the Polymerase Chain Reaction, but when the lights go down and the slides come on, well . . .

John Martin, President of the European Society for Clinical Investigation, said in Nature, *"His only slides (or what he called his art) were photographs he had taken of naked women with colored lights projected upon their bodies. He accused science of being universally corrupt with widespread falsification of data to obtain grants. Finally he impugned the personal honesty of several named scientists working in the HIV field. . . . The council of the European Society for Clinical Investigation will not be inviting Dr. Mullis to further meetings."*

Really, do you need this in your community? Of course not.

And now, for a limited time only you can be assured that Dr. Mullis will not ever lecture at your society, school, research lab, etc.

You personally . . . and confidentially . . . can assure it.

Call now at (my phone number) and ask for (my beautiful assistant). Have your Visa or Mastercard card ready. Prevention rates begin at $500 per year guaranteed and are progressive with the size and sensitivity of your organization. You may request personal anonymity, or for $79.95 plus shipping we will send you a Special Service Award embossed with your name and a special inscription commending your judgment, *foresight, and unselfish devotion to your community. Custom inscriptions are a little extra but can be especially commemorative.*

Think about honoring your boss or one of your associates by taking advantage of our special "Help a Friend Stop Mullis" offer. Call for details. Don't delay. Only one offer of complete protection per year can be extended to any single organization. Be first. Be smart. Be safe.

Recently, Glaxo Pharmaceuticals found it necessary to send Dr. Mullis a check for $6,048.00 simply to prevent him from speaking at their annual Chemistry and Medicine at the Frontiers Conference in Chapel Hill, N.C. No one at Glaxo had seen fit to acquire protection from a Mullis seminar, and haplessly, Dr. Ross Johnson, now no longer with Glaxo, had invited him.

I must report that the response to this offer has been underwhelming. Neiman-Marcus has not chosen to include it in their famed Christmas catalogue. So I have continued to speak out to any forum when I have been given the opportunity.

It is not too late, however. If you would like to give the gift of my silence to an individual or an organization, all reasonable offers will be accepted.

NOTE: This offer is not open to family members or employees of Kary Mullis, who are doomed to have to listen to what I say.

20

AM I A MACHINE?

D<small>O</small> I <small>MAKE</small> things happen? Or do my arms wave sometimes due to the rigid, absolute, causal connections between consecutive moments in my brain, and nerves and muscles? Are my sensations of being here as an active part of this thing just another part of that causal chain? Who cares?

At night I care. It gets spooky thinking that maybe I am a machine with no feelings that matter.

Am I a machine? Are my future states all plotted out for me by physical laws? If they aren't, then at what point does something happen that is not determined by the laws of force and matter and quantum mechanics, and General Dynamics, and General Motors, and the surgeon general, and TRW? In other words, is there something that I can do called "exercising my will" that is outside of time and space, where everything otherwise has to behave according to equations? Do I do it constantly, or only once a week? Am I a machine or do I have sovereignty? Autonomy? Is freedom "just another word for nothing left to lose," or does it mean that there is a ghostlike phenomenon associated with my body right now that can move things but cannot be moved? Sound unlikely?

There are some nice buildings in America and some even nicer ones in Europe where men, wearing fancy robes, surrounded by great art and nice music, will tell you that it's not at all fanciful; in fact, that it is absolutely true.

Consider the following. I know—before anyone else could—when I am about to stand up in some place and scream, quoting e. e. cummings, "I will not kiss your fucking flag!" Everybody else would be surprised, but I would have known it before they did. I would have the impression that I caused it to happen. It's my mouth. Or is it?

Imagine a deserted building in Texas. It was originally going to house the supercollider before the money dried up. It now houses a massive computer, and you are being wheeled up to it. Some serious guys in long white coats are plugging you in. The electrodes on your head are revealing to the computer everything that is going on in your brain.

The computer prints out, in lurid color, a complete and accurate description of what you are about to do. Now, only you and the computer would know just immediately prior to when you started to loudly quote cummings. Would you be worried?

What if the computer got way ahead of you and could have already written a few hundred pages about how you were going to respond to some simple movies it was going to show you? And the guys in the lab coats had already read it and were laughing and having coffee before you even did it? What if it was able to generate a videotape of exactly how you would react to something long before you did? Would you be nervous? Would you wonder who you were?

I would. Especially if it were dark. Is this machine responsible for you? Are you?

In a society where there are maybe too many people, it is reasonable to weed out the nasty ones. Killers, robbers, rapists, lawyers, editors, not necessarily because they are guilty for what they do but just because they do it and it hurts us. Guilt is sort of the theme of this chapter. What is guilt?

Not the feeling of it. We all know that. But guilt itself. Guilty enough to be administered a lethal injection by the state kind of thing? What is that? Am I a machine, just watching myself do things and feeling responsible, because I get a preview of what I am about to do, by being able to see into the future states of my brain? Or am I an eternal being, separated completely from time and space? Or am I something else?

If you live in a village under a steep mountain slope, and you have village meetings once a month, eventually someone will bring up the notion that a wall should be built to keep rocks from falling on roofs. No insult intended to the rocks. The rocks would understand, if they could think, that their downward motion should be impeded. After all, there are women and children in the village. And property. And men. The rocks hurt people and should be stopped.

What is the difference between you—hooked up to a computer that can predict your future before you can imagine it— and a rock falling down a mountain with a camera monitoring its descent, the same computer predicting its future course? What's the difference between you and a little car tripping around on Mars with a bunch of grownup high school boys— like gods far away—controlling it? Now where do you fit in? Where do I? Who's running the show? Is anyone watching the store? Who is to be punished? Restrained? Destroyed? The rocks falling on the village? Sure, the rocks. Who else?

These are not jokes from a B movie. These are issues that

our culture is presently dealing with, although not well, and usually at what sounds like a more dignified level. It comes across as something like this: "Was this defendant psychologically capable of knowing wrong from right when he smashed in our roof?" That kind of shit. And excuse me, but I can't help thinking that that is pure shit. We can't seem to frame these issues properly. Maybe the reason is that we are toy cars. Or maybe the reason we can't deal with the realities of tumbling boulders is that we are not toy cars. Maybe we are evolutionary beings, grown up by random chance, out of nothing but clay, on a hostile planet. We have no idea where we came from, and a sad lack of imagination.

Maybe we are ourselves just helpless rocks.

I, on the other hand, have a memory and I have free will. I assume you do, too.

"Oh yes, you do," comes a voice out of my dark bedroom closet while Nancy is out of town and I'm alone. A squeaky little voice followed by a light tapping sound like a keyboard.

It speaks. "Maybe your memory is just a feeling you get from the fact that your brain is floating through time and it can peer back into past states of itself. The seeing isn't all that good, it gets worse with age, but it can see back into its past states. It can see what its eyes once saw, or hear what its ears once heard. It can't see into *my* past states, though. Only into its own. It feels a private relationship to itself. Almost a feeling of ownership after a while. Know what I mean?"

"Yes. That sounds reasonable." I pull the covers over me. Italian sheets—400 threads to the inch. Nancy buys them from David at the Golden Goose in Mendocino and I love them, but they aren't a comfort to me tonight, hearing this voice in my closet.

"It is reasonable," squeaks the voice, and the tapping stops. "And think about this. That feeling you get that you have control of yourself? That's just caused by the time symmetrical reflection of what you call your memory. Know what I mean?"

"What?"

"That's caused by the fact that your brain, which is only a wave phenomenon passing through time, can not only see into its past, but also peer into its future states. See what I mean?"

"What about that computer they can hook up on my brain and figure out what I'm going to do before I ever do it?"

"What about it?"

"I don't know. I'm too tired. Go away."

"Did you turn off the printer?"

"Go away!"

21

PROFESSIONAL BIOCHEMISTRY

THE REASONS I took up biochemistry as a profession were simple. With Mercury and Mars in conjunction in Sagittarius, I was not going to specialize in something well-defined and manageable. I didn't think of myself as a worker, or a specialist. I thought of myself as a man of deep science with a Gemini moon in my face and the cold, red winds of Mars in my hair. I wanted to see reality, if possible, and my Capricorn sun felt a strong need to make a living.

The choice was going to be between the study of my body in great detail or the study of everything else. Either I would become a biochemist or an astrophysicist. I lingered over both, but my body won out.

The government was paying for graduate school in both disciplines at the time. It was 1966. I suspected that as soon as the Russians were no longer a threat, funding for the study of things in outer space that nobody will ever touch would drop off a congressman's list of the essentials. The study of human bodies, and the things that go wrong with them, I surmised, would continue to be funded. I didn't realize the Russians would eat it so soon, or that biotechnology would bring in so much private funding, but overall, I was right.

There was also the social excuse for choosing biochemistry. The universe sounds pretty universal, but try discussing it at a party with a twenty-two-year-old woman who never thought about neutral kaon decay rates as her trip. Then talk to the same woman about why ethylamino derivatives of safrole like MDA will make you want to take off your clothes and feel warm and cuddly for about eight hours. Even though she has never thought of catecholamines as her trip, she might be curious about this. A strong social impulse will lead you away from astrophysics and toward biochemistry. It did me.

Biochemistry was more fun. It still is. I don't go to parties for the twenty-two-year-old girls anymore, but nothing is more fun or interesting to me than human bodies. I am one. I want my eyes to keep focusing, my heart to keep beating, and that thrilling sexual function my body engages in to keep working night after night.

I like to know about those things and all the diets and drugs we ingest to keep it all working. Furthermore, I like to fool around with it. I liked to make chemicals in the 1960s that had effects on my mind. I like to make chemicals in the 1990s that have serious effects on anything alive.

I like to make chemicals that could turn a sponge into a gold miner. A happy little creature that filtered water like a normal sponge might be endowed with a voracious appetite for filtering out the gold that washed down the Sacramento River out of them thar hills. I like to make chemicals that might help heal a spinal cord that had been crushed by its owner's motorcycle. I'd like to cure diabetes. This is what biochemists do.

Down below the limits of our vision, there is a level of physical organization where the parts are like little machines and conveyor belts and tables that hold the machines and partitions between them. There are things with shapes like Tinkertoys that go through machines and get turned into useful shapes like springs and sockets and drops of cement that hold things together or fill in the cracks. And none of this stuff is beyond our understanding.

Most of it can be figured out in terms of what things stick to each other, or twist or push one another. The things that become other things by being stuck together or cut in half, the things that are made in one place and then have to be moved to another place, and the things that will only function when fitted with other things. Stuff like that. Biochemistry is a lot like mechanical engineering or auto mechanics—only you can't see the parts with your eyes and you don't get your knuckles bloody or your fingernails greasy. Sometimes you get poisons on you, but you never lick your fingers.

The most important principle is that living systems are modular. They are collections of cells, and the cells are collections of parts, and we have a rapidly growing familiarity with the nature of the parts and the ability to make them. We have names for them and pictures of how they would look to our eyes if they weren't so small.

Sometimes the shapes are complex and there are so many parts that the processes involved in disease and health seem endless. Spinal cord injury is a little like that. The human spinal cord contains millions of tiny tracts, and thinking about trying to repair it sometimes seems like trying to repair the ion drive on a crashed flying saucer. But there are those of us who seriously work on that. I work with a team jointly sponsored by Immune Response and Vyrex corporations. We're looking seriously at spinal cord injuries. We're starting at the bottom, trying to catalogue all the genes that might be involved. It's slow, but there's a quarter million people in America alone who can't feel or move their lower body.

I like to work on something where the defect is obvious and the solution is simple. Insulin-dependent diabetes mellitus is one of those diseases. I'm trying to organize an effort to cure that. If you have IDDM, then you have to take a chemical called insulin because the conveyor belts that move sugar into your cells when it's needed don't function without it and you can't make it.

Insulin is normally made by a special little shop in the pancreas that goes by the charming name of the Isles of Langerhans. If insulin were all those cells made, then insulin would itself be a cure for diabetes, but that's not the case. It helps, but you need those cells. In people with IDDM, something kills them. We now know what that is. It's another set of cells, called T-cells, whose job is to kill other cells that are screwed up for some reason—like tumors, or cells infected with a virus. Like other hired killers, they have to be carefully controlled. In the case of IDDM, one set of T-cells has gone wild. In the summer of 1997, there was an article in a scientific magazine called *Cell* that described why this particular set of T-cells, called Vbeta7-T-cells, grows out of control in people with IDDM. The cells have accidentally started to make a chemical that not only allows them to reproduce themselves frequently, which they do, but also allows them to arm themselves. This particular set of T-cells would not normally be allowed to reproduce, or be armed, because they happen to have the noxious property of attaching themselves to the Isles of Langerhans. When they do, and they are armed, they release acids, peroxides, and biological warfare agents that the quaint and unarmed cells of the Isles of Langerhans have no defenses against. They die. It takes a long time, which is why IDDM is a slowly progressive disease.

The chemical that the Vbeta7-T-cells have started to make that drives them to do this damage is a protein that unfortunately functions as an adapter between the T-cells and their nominal bosses, the antigen presenting cells. Don't think this is any more complicated than the nose on your face, because it isn't. This is an easy problem that can be solved as soon as a few biochemists are directed to work on it.

The antigen presenting cells are tethered to the T-cells by this adapter protein that shouldn't be there. Thus tethered, the antigen presenting cells give the T-cells permission to reproduce and to go crazy. The problem is the adapter protein itself.

Biochemists in 1998 know how to deal with proteins, just as the military knows how to deal with underground bunkers. All we have to do is come up with another protein or some other chemical that can take that mother out. And if there is anything that biochemists can do, that's it. There are several possible ways to do it and probably any one of them will work.

So why isn't IDDM already cured? The paper in *Cell* has been out for about a year. It is a matter of convincing people with *money* to pay people with *skills* to focus on this project. Because I am an older, more experienced biochemist, that job of organizing and promoting large-scale efforts falls on me. I'm not all that good at it. Being clever in the lab doesn't make me clever in the boardroom. It's the same in every technical profession. The older people who worked their way up being an engineer or a scientist now have to try to organize people. It is not so hard for them to deal with the technical people under them; they were there once themselves. But they also have to interact with people out of their field of expertise—business and financial people, patent lawyers, public relations people, regulatory people, marketing experts, and so on. It's a challenge.

Wish me luck.

22

THE AGE OF
CHICKEN LITTLE

I WAS RIDING my bicycle up Mount Soledad in La Jolla
this morning, and I was huffing and puffing a little more
than my fair share. I was giving off water vapor and carbon
dioxide—both greenhouse gases—like an animal trying to
escape a predator.

Greenhouse gas means that light streaming in from the
sun—which is equivalent in intensity to the light from a 100-
watt bulb sitting directly over every square foot of the Earth in
the daytime—will pass right through water and carbon dioxide on the way down. That light hits the Earth and warms the
ground. The ground, being warmed, tries to cool off by making
another kind of light called infra-red. This is more like the
light that comes from those red lightbulbs in hotel bathrooms.
The infra-red coming off the warmed Earth radiates back
toward space, but it doesn't escape. It will not pass through
water vapor or carbon dioxide. It gets absorbed, and that heats
up the atmosphere. It is called the greenhouse effect and is
why much of the Earth is warm enough for T-shirts and shorts.
It's also why a sweaty bicyclist sailing back down Mount
Soledad, releasing water vapor into the air, would not attract
polar bears.

I wondered on the way down whether the United Nations

Intergovernmental Panel on Climate Change was watching me from some satellite, recording my blatant and unnecessary contribution to Global Warming. With a budget of over $1 billion a year, who knows what those international bureaucratic bastards are up to?

Okay, maybe I am a little overzealous here about the IPCC. But they are causing us a lot of trouble and we are happily paying for it the same way we paid for the Inquisition some years ago, when another international bureaucracy called the Catholic Church got out of hand.

We have been had again, and grossly misinformed. And the more we pay these parasites, the longer they will be in business and the more damage they can do in the name of saving us from ourselves.

The Catholics and their associated henchmen, the revisionist Christians, fixed our dues at 10 percent of our income. The climate control cartel—which includes everybody who can charge us for measuring a climatic variable and claim that it is changing in any way, shape, or form—is now spending more of the world's resources than we used to allocate to the much more realistic threat that someone might blow up the world without the threat of fierce retaliation by other concerned parties. It was mad, of course, and that's what they called it—MAD, or Mutual Assured Destruction.

But this business of intergovernmental panels on climate change is not just mad—it's embarrassing. Furthermore, it smacks of what the Greeks used to call hubris when one of their number decided he, and not the gods, could control his own life, the weather, or something equally impossible to control.

If it were just the embarrassment or the mortal sin of hubris involved, I don't think I would get upset about it. Everybody needs a job. But people should only be paid for doing things that benefit the people paying them. No one has ever been able to predict long-term weather better than a tossed coin. Why do we continue to pay a vast cadre of scientists and bureaucrats who pretend to speak for the Planet?

They claim that we can change the world forever—and they are willing to tell us exactly how. The U.S. Weather Service has gotten a little more conservative about saying things about the future. They won't even make ninety-day forecasts any more. They used to do that, but after 1988 they ceased the practice because they noticed that a coin flipped was cheaper than a cadre of computer scientists and just as accurate.

People are jerked about almost monthly by new announcements by spokesmen for various government agencies and research groups sponsored by government funds. They tell us that every time we start our cars we contribute to greenhouse gases. Every time we vent Freon from a refrigerator, air conditioner, or spray can into the atmosphere we are destroying the ozone layer and contributing to the worldwide incineration of all life. It makes no sense, in the light of the climatic history of the world, to talk about catastrophic changes in the weather being caused by human activities.

What happened in the 1980s? We have brought something down on ourselves as expensive, although not quite as brutal, as a world war. Did everybody forget that we were just big ants? Did somebody convince us that just because most of our religions had lost their appeal, we ourselves were suddenly gods? That we were now the masters of the planet and the guardians of the status quo? That the precise climatic conditions that happen to exist on the Earth today in the Holy Twentieth Century, the Climatic Century of 001, the first year of human domination of all of Earth, should be here forever, *in secula seculorum? All the good species are here now.* None shall perish and no new ones are welcome. Biology is no longer allowed: the Environmental Protection Agency and the Intergovernmental Panel on Climate Change are now in charge. *Evolution is over.*

I recall a cartoon. A caveman is raging in front of his cave glaring up at a flash of lightning and pointing an accusing finger toward his mate and the fire burning in the mouth of the cave, "It didn't used to do that before you started making those things."

The future of the Earth has got nothing to do with the creatures that live clustered along the shores of its great bodies of water. We are just here for the ride. And the ride is not smooth. It never has been smooth.

The world that the Vikings sailed out into a thousand years ago was warmer by far than it is today. Since then it has gotten colder. It even got colder last century. It didn't do so in response to the Viking ships or the Spanish horses dropping manure on the California poppies. It got colder all over the planet and drier on the West Coast of the United States for reasons that only the planets and the sun can be held accountable. It pissed off the Spanish who were trying to civilize the Indians in California as slaves, growing crops for the missions while having their souls saved. The Indians laughed that the white man's god could not provide rain for the white man's foolish crops. It got colder and drier because angles and distances of Earth and our major heat source changed—things that neither the Vikings nor the Spanish could measure and surely did not affect.

When I start feeling a bit of a chill in my cabin in Mendocino, I don't worry too much about greenhouse gases or ozone. I move toward the stove, and invariably I get warmer. We can't position our planet relative to the sun just exactly to our liking. We can't make sure that the average temperature in San Diego for the next thousand years will be a comfortable 68 degrees Fahrenheit. But we can stop worrying about the minor things in the atmosphere that we don't understand. We can stop worrying about whether we can control it because we don't have anything to do with it. It just plain isn't stable. It may get colder in the coming centuries—it may get hotter.

About 11,500 years ago the surface temperature of the Earth began to warm. The glacial period that had lasted for about 100,000 years was ending. It had been about 20 degrees Centigrade colder. For Fahrenheit people, that's 36. The present interglacial period is a vacation for *Homo sapiens*. We can sit out on the front porch of the cave on a lawn chair and enjoy the sunset. We can even mow the lawn instead of shoveling snow.

There was another interglacial period, the Eemian, which ended about 120,000 years ago. From the tree ring data found and the ice cores they drill in Antarctica, that also seemed like a pretty nice break in the weather. Of course, I prefer to surf than ski. As we go into the next millennium, all the solid facts look like I'm out of luck. We are headed back into another glacial period, which is a more common climate on Earth than the relative warmth we are enjoying now. So who's bitching about global warming? Is it the skiers? It's not the surfers.

The global warmers—the climate simulation programmers, the so-called general circulation modelers, the computer jocks who hardly go outside even on nice days—write the programs for their bosses at IPCC. They predict that global warming is coming and our emissions are to blame. They do that to keep us worried about our role in the whole thing. If we aren't worried and guilty, we might not pay their salaries. It's that simple.

If we had sailed into here in space ships and the physical history of the place was that the climate had always been the same, then we might reasonably think that there was an amazing delicate balance on the Earth that we should not upset, if for no other reason, just to show a little respect. Maybe we could justify hiring experts or priests to help us.

But that is not at all what happened. We evolved here, and we evolved in the midst of some pretty serious climatic changes. They were serious enough so that millions of years and extinctions later we can still see the effects of the changes and give names like "carboniferous" and "cretaceous" and "Eemian" to the very different climatic epochs because they were different. There is no reason to think things are going to stay the same now—with or without us.

The Earth is a massive thing sailing majestically around the solar system feeling the gravity of the sun and the planets and their moons, and the asteroids. It takes a hit now and then from strays like the Shoemaker-Levy vandals that Jupiter entertained last year. It submits to the gravitation of the Moon, and it answers with the tides. It bends its ionosphere to the solar wind and it feels the massive gravity of Jupiter pulling it slightly. But "Old Blue and Green" endures. It does not bow to humans or ants.

The temperature of the Earth is due to the size and shape of the orbit that it follows around the sun, the angle that its rotational axis is tilted to its orbit, the length of its days, the radioactive decay and residual gravitational heat deep below the crust, and the elements that were here from the beginning, and God knows what else, but not us.

We are a thin layer of moss on a huge rock. We are a little biologic phenomenon that makes words and thoughts and babies, but we don't even tickle the soles of the feet of our planet. We pick and dig around on the outer skin and mark it for ourselves in little squares. We look out at the stars and think they, too, are for us. In spite of the grandeur that we see, we still entertain the most unusual ideas about ourselves.

Is it because we are afraid of the dark or death that we have to puff ourselves up and be kings of creation, masters of everything, protectors of the planet? How can we pretend to be masters when our flashlights are always running out of batteries and leaving us in the dark? What about the fear that the sudden lack of human eyesight causes? Does that seem like something a planetary master would feel? If all the energy modes on the planet are summed, with all the information on each of them, most of it is still indecipherable to us and most of the energy modes over which it is conveyed are still invisible to us. There's nothing significant missing when the flashlight goes out. If it's dark without it, then it was dark with it. We are only watching a couple of channels. There are a million.

The vast majority of the world is invisible to our eyes regardless of the brightest of our lights, and we can't hear more than a tiny bit of the sound of it with our ears, and we can't feel the subtle textures of it with our fingers. Even with all our instruments, long tubes on mountains, and a Hubble telescope in space, we are blind to the myriad of complex energies that are whirling and vibrating and clattering all around us day and night, year after year, millennium after millennium.

The appropriate demeanor for a human is to feel lucky that he is alive and to humble himself in the face of the immensity of things and have a beer. Relax. Welcome to Earth. It's a little confusing at first. That's why you have to come back over and over again before you learn to really enjoy yourself.

The sky is not falling.

ACKNOWLEDGMENTS

I have always wanted to write books. I've tried before; they are always half finished or less.

After we had been together for some time and it was clear that I was a scientist, my third wife, Cynthia, confided to me that she had always wanted to marry a writer. She would read Agatha Christie to me at night, and I eventually dropped science for a while to write. The immediate result was a story, a payment to me of $120 from a magazine called *Medical Dimensions,* and a day job working at a restaurant. The final result was that I went back to work as a scientist and won the Nobel Prize.

I gratefully acknowledge the help that receiving that prize has been in my life. There is money involved, but what is more lasting is that once you have been given that accolade, no door in the world will fail to open for you at least once. It is a free pass for the rest of your life. I herein thank the members of the Nobel Committee for doing the right thing and giving it to me while I was young enough to enjoy it. It is true, I am a loose cannon on deck, and you were taking a chance that I would disgrace a fine institution, but I think you will not regret what you did.

And thanks to Pam Ingate, my secretary and surfing-dude friend, who has always been devoted and tireless, and a bright, happy place in my life.

I would also like to thank Toni Cosentino, who, after the O.J. trial and years after Cynthia had raised the issue, seconded the motion. Toni suggested that I write a book, and by car phone, on a trip between L.A. and Santa Barbara, she introduced me to her agent friend, Frank Weimann. Frank introduced me to David Fisher, who co-authored the first version of this book.

I began to rewrite my book when I realized that you can't have somebody else buy your clothes for you unless you are either totally relaxed about what you wear or don't have your own taste. The same applies to a book about dancing naked anywhere.

Nancy Cosgrove convinced me that I could make this book come from within me. She liked my writing. And then Nancy, the painter I was already falling in love with, became Nancy the editor, who would see me through every page and all the traumas that arise from having an incomplete manuscript that is six months late.

And almost every day Nancy talked with Altie Karper, the managing editor at Pantheon, and Altie began to trust us. Her guidance and faith and patience are very greatly appreciated. Altie and Erroll McDonald, vice president and executive editor at Pantheon, then had to convince everyone else at Random House to trust us also, and I guess it worked. Thanks, Altie. Thanks, Erroll. We finally got it done, and we're happy with it.

And then Nancy and I got married. The book is done and she has time again to paint.

INDEX